# HOW TO "HOP UP" CHEVROLET & GMC 6-CYLINDER ENGINES

By ROGER HUNTINGTON
MEMBER, SOCIETY OF AUTOMOTIVE ENGINEERS





FISHER er BOOKS™
4239 W. Ina Road, Suite 101
Tucson, AZ 85741
520-744-6110

# *History Revisited*

Forty-seven years later I am pleased to reprint Roger Huntington's *How to Hop Up Chevrolet & GMC 6 cylinder engines,* originally published in 1951 by Floyd Clymer in Los Angeles. Treasured by enthusiasts, by the 1990s these books had become almost impossible to find and good copies were selling for $50 at swap meets.

I personally knew Roger Huntington and always appreciated his insightful articles on automobile design and especially on automotive performance.

Although Roger Huntington was a paraplegic, he never let that stop him in his search for automotive knowledge. You just never knew when you would see Roger in his wheelchair–at drag races, at press introductions at GM, Ford or Chrysler, out "test driving" as he rode along and got impressions of how a new car handled and performed. Then he would write about what he learned and how he felt about the car. His byline appeared in almost every automotive magazine of the day. Roger Huntington's was synonymous with then-current knowledge about high performance.

I know you will enjoy reliving history as you turn the pages of this automotive performance classic.

Thanks to Keith Young for the beautiful cover photo of one of his Wayne Chevrolet engines. This particular engine was chosen for display in the Petersen Automotive Museum in Los Angeles during 1996-1997.

Bill Fisher
September 15, 1997

# Announcement

This is the second book we are offering (following Mr. Huntington's "Souping the Stock Engine"), which has been compiled to meet increasing demands for specific information concerning the "Souping" or hopping-up of certain stock production automotive engines. It deals with Chevrolet and GMC engines in particular – both becoming very popular for use in racing, hot rod, and competition cars.

Basically the same in organization as the previous books, this one surveys the various speed tuning theories, the range and availability of speed equipment for these power units, and pays special attention to the cost economies involved – how to get the most for your investment in rebuilding Chevrolet or GMC engines. Mention is made of several new brands of speed equipment on the market, about which it may well be profitable for the reader seriously contemplating this work to inquire. The advice offered on planning and computing performance results also will be beneficial, as experience in the field of speed tuning has shown.

That men working with these engines have faith in their potential performance is evidenced by the fact that both Chevrolet- and GMC-equipped cars were entered in the 1951 Indianapolis Race. Although neither of them succeeded in qualifying for the race, their respective lap speeds of near-125 mph is sufficient proof that these engines can be made to perform at high RPM. Consistent records at drag-races and time trials in Southern California (and the S.C.T.A. National Hot Rod Trials at Bonneville Salt Flats) is further proof.

Definite reference is made herein only to automotive installations, but the information applies also to marine, aircraft, or any other situation where *increased power output* is desired.

In our opinion, this new book is the best available theoretical treatment of this subject, and will become a standard text in the field of speed and power. We hope you enjoy it, and that you will derive real benefit from its contents.

Floyd Clymer

Publisher

*A view inside Nicson Engineering Co., Los Angeles, one of several successful Chevrolet and GMC speed equipment manufacturers on the Pacific Coast.*

# Preface

"...we do not subscribe to the idea of reworking or modifying Chevrolet engines for extra high output or for racing purposes. We do not feel that Chevrolet has anything to gain, either from this activity or from books published in that connection. Accordingly, we will be unable..."

This little tidbit is quoted from a letter to the author from the Chevrolet front office in Detroit. It pretty much sums up General Motors' official opinion of what we write in these pages.

And who are we to say they're right or wrong? Any company that sells over a million cars a year has something on the ball. But at this point, we dare say that Chevrolet's opposition to any form of competition involving their product is *not* one reason why they sell a million cars a year. Competition and the engine and chassis development that stem from "souping" can't help but benefit the American automobile and the American motorist in general.

There's John Smith, obscure draftsman in Chevrolet's engineering department. He subscribes to front office theories on company time, but he's running a Nicson dual, an Iskenderian cam, and a few other "items" in his Chevy on his own time—(and he's the only guy in the parking lot who can burn rubber in second gear!). Smith is a better automotive engineer because of his first-hand experiments with his souped-up Chevy.

And so it goes. Souping is a great thing all around if you don't abuse it. But the point is this: Your stock car—Chevrolet or otherwise—never was designed for the performance you're trying to get out of it. Remember this if the bearings go out after 5,000 miles, or if you crack a piston from detonation, or if you tear out the transmission on one of your "drags", or if the whole rig begins to wander a bit at 100 mph!

The minute you so much as stick a high-lift rocker arm on your Chevy, you've burned a bridge behind you—the bridge between you and the Chevrolet public relations department. They just won't stand behind a hot rod for more reasons than one. In other words, don't expect too much of a U. S. stock car under souped conditions. Use your head instead to improve that car from bumper to bumper. That's the true spirit of "hot-rodding".

In this book, we're going to try to guide you in souping the Chevrolet 6-cylinder engine for high-performance vehicles. We can't hope to touch on every part and piece of equipment down to the last bolt, nor will space permit us to list step-by-step instructions on every operation. We're going to stress instead the *planning* side of the job, with specific instructions in this category. General instructions will be given on procedure. In the Appendix is a section on souping the popular GMC truck engines for use in Chevrolet cars and for racing.

ROGER HUNTINGTON

# Table of Contents

## Acknowledgment

We desire to thank the following firms and persons which assisted in supplying data on their products, photographs, and general information used in this book:

Wayne Manufacturing Company
Nicson Engineering Company
JE Engineering Company
Ed Iskendarian
Edelbrock Equipment Company
Italmeccanica, Inc.
Besasie Engineering Company
Newhouse Automotive Industries
Lee's Speed Shop
Wolfer Corp.
Hilborn-Travers Engineering Co.
SpeedOmotive
Chet Herbert
McGurk Engineering Company
Witteman Company
Tattersfield Company
Edmunds Equipment Company
O'dell-Shields Studios
Automotive Industries
Asher Lee

*Fig. 1-1. Johnny Hartman's famous No. 59, the first really hot competition car to use the 6-cylinder Chevy block. The engine was built by Wayne Horning and developed 220 hp.*

# CHAPTER 1

## THE WHY AND THE WHEREFORE

NOT so long ago — only a few years, in fact — 200 horsepower from a Chevrolet block was just a wild dream of a young Los Angeles mechanic!

In May, 1948, after several years of experiment, Wayne Horning threw the book at that lowly stock mill and came up with an astonishing 220 hp from a Class B piston displacement of 248 cu. in. This high-revving fireball went into the now-famous No. 59 *"Hartman Torsion Bar Spl."*, a glorified hot rod that immediately began showing Fords and Mercs the short way around on California tracks — and chalked up a sizzling 140.18 mph on the dry lakes.

So after a decade of packing around a lower end that "had the stuff in it", the 6-cylinder Chevy finally flexed its muscles, and thus started the soupers' stampede toward this beautifully-designed General Motors production engine. Today they tell us you can get 1 hp for every 3½ lbs. of engine weight in the 150-hp range for less money with the Chevy than with any other engine in the world!

So here we are with a big problem on our hands: How to soup up that Chevrolet block to get the most power for the least cost, and how to plan for the best possible performance and service under our particular "souping conditions". Now that all seems pretty obvious. But is it? Let's investigate these basic aims a little more closely before we haul out the wrenches and calipers.

### WHY SOUP?

That's not a silly question. Here's Leadfoot Louie, our moron hot-rodder, recently turned from the V8 to the Chevy block, who's busily engaged souping his engine to get (and we quote) — "200 hp, 25 miles per gallon, 100,000 miles between overhauls, and smooth, flexible performance at all speeds." What a rude surprise Louie's going to get when he adds up his operating and repair bills, when he steps down on it at 15 mph in high gear, or when he tries to keep it going through town on a cold morning (if he can start it)!

No, you can't have everything. We're after just *one thing* when we soup up a stock engine — maximum horsepower at all speeds. Even this is a pretty big order. But the point is, don't expect pennies from heaven in your souping — things like economy, easy starting, silent running, smoothness, and flexibility When you're after power, these items just don't follow hand-in-hand. In other words, your souped engine is great for "hot" transportation — not so hot for transportation *in general.*

Which brings us to another important question: If all we're after is 150 or 160 hp, why not buy a new reconditioned Packard, Lincoln, or Cadillac engine that will give this much in stock form? Then we wouldn't have to lay out a lot of money for special parts, and we could cash in on those extra dividends such as economy, silence and smoothness.

This question gets us right to the heart of this souping business. It's all

*Fig. 1-2. The Chevrolet-powered "Johnson Spl.", entered for the 1951 Indianapolis race. The Wayne-Chevy engine developed 265 hp at 5000 rpm on methanol fuel. Of several semi-stock engines entered for the "500", this car had the highest lap speed — turning a consistent 125 mph.*

*Fig. 1-3. Grand Prix maestro, Juan Fangio, in his Wayne-equipped Chevrolet racer in South America. He won the Buenos Aires G.P. last winter.*

a matter of engine size and weight. To get the maximum possible speed and acceleration out of any vehicle on a given HP — be it a car, boat, motorcycle, or airplane — we want the lowest possible gross weight and size. This, of course, implies a light, compact engine.

Large passenger car engines such as the Lincoln and Packard don't fill the bill; they're bulky, and weigh in the 700-900 lb. range. The little Chevy 6, on the other hand, is compact and weighs only 575 lbs. Special souping equipment doesn't appreciably alter this total weight. In other words, we soup to get a powerful, *small* engine, not just a powerful engine. (In all fairness, we must admit that, if engineers go much farther with stock engines — as witness the 700-lb. 180-hp Chrysler V8 — they're going to make pikers of all of us who spend our time and money souping a Ford or Chevy!)

While still on this subject of why we soup a stock engine, we might mention one other good reason: Because we can't *afford* a *really* hot engine. That sounds funny in this day of 200-hp Fords and Chevys, but think about it: For something between $2,000 and $10,000, we could go and buy an out-and-out racing engine that would kick anything from 200 to 400 hp for a weight in the neighborhood of 500 lbs. — on pump gas!

But not one in a thousand can put up that much money — hence this souping business. We might sum it all up this way: If we're going to get 150-250 hp for an engine weight of 500-700 lbs., and for a total investment of well under $1,000, we have no choice but to turn to the small stock block.

*Fig. 1-4. Three beautiful Chevy engines being crated at the Wayne Mfg. Co. for shipment to South America, to be used for stock car road racing. These units are built with 8½:1 compression ratio to burn the low-octane pump gas available in the Latin countries. Price, around $1,050 each.*

*Fig. 1-5. The Chevy takes to the water. The Wayne engine installation in "Lil Injun" Class E racing runabout.*

*Fig. 1-6. Operation shoehorn. A Wayne-Chevrolet engine installed in a '34 Ford roadster; note that the firewall has been deeply recessed to accommodate extra length of the Chevy block.*

## PITY THE POCKETBOOK

Souping is not a poor man's game. And the fact that the Chevy is one of the most inexpensive engines on the market doesn't mean that a beggar can pour speed parts to it. As a matter of fact, the Chevy is in an odd position in this respect as compared to its arch souping rival, the Ford-Merc V8. With the V8, a man with very little money can have his heads milled for $10 and take his first significant souping step; with the Chevy it isn't quite so simple and you can't do anything that will help much for less than $30.

But on the other hand, as you go up the souping scale toward more and more power, your cost per HP drops below that of the V8. It's a funny situation, and it's all due to the overhead valves. Just remember that the pauper will have a harder time getting started on the Chevy block than he would with the V8.

Then there's the matter of labor. A lot of people, for one reason or another, can't do a bit of their own engine work such as disassembly, adjusting valve clearance, etc. It's going to cost these fellows just about $3 per man hour to have this work done for them — and this must be taken into account in their cost estimates. In general, it will run something around $70 in labor to pull the Chevy engine, tear it down, assemble, and replace it in the car. Labor costs for the smaller jobs will scale down from this. In other words, for the boy with limited funds who can't do his own work, labor costs will be a very important item, and may prevent tearing the engine down in the early stages.

For the rodder who has the ability, equipment and time to do his own engine work, we have a different situation. He can do wonders with $100, and he will be able to build up his super track engine for perhaps $200 less than the "white collar" souper. So the factor of costs will be a major item in any souping plan.

We might sum up the matter of costs this way: If you can't do any of your own engine work, it will take $50 to do much with the Chevy, and perhaps $600 to get upwards of 200 hp on pump gas. If you can do some of your own work, you might do business for $25 at the bottom end and $400 at the top.

These are hard, cold facts. And the fact that this souping business is something that gets into your blood makes the cost factor even more critical. If you've ever had a *fast* vehicle under you, you know it's hard to be satisfied until it is just a little faster. That means more money. It's something like a drunkard who wants his liquid cheer whether Junior gets the new pair of shoes or not. Similarly with souping — some can afford it, some can't. It's well worth some sacrifices, but don't lose your head. If you can't eventually see a couple of hundred dollars in special parts for your Chevy, don't expect miracles on the road.

So that is what this game is all about. It's a great life. It's certain that any sporting thrill, be it running 85 yards through Notre Dame for a touchdown, beating Sam Snead on the golf course, winning a World Series with a home or riding the Kentucky Derby winner, will have to go some to equal the thrill you get when you build up a car with your own hands, gun it, then feel it try to pull out from under you and run up to 120 mph.

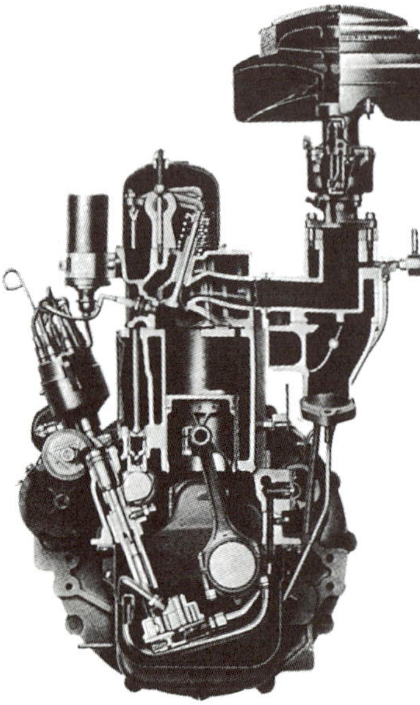

*Fig. 2-1. (Left) — Front cross-section of the basic Chevrolet engine (Powerglide).*

*Fig. 2-2. (Below) — Side cross-section of the late standard 216-cu. in. Chevrolet engine.*

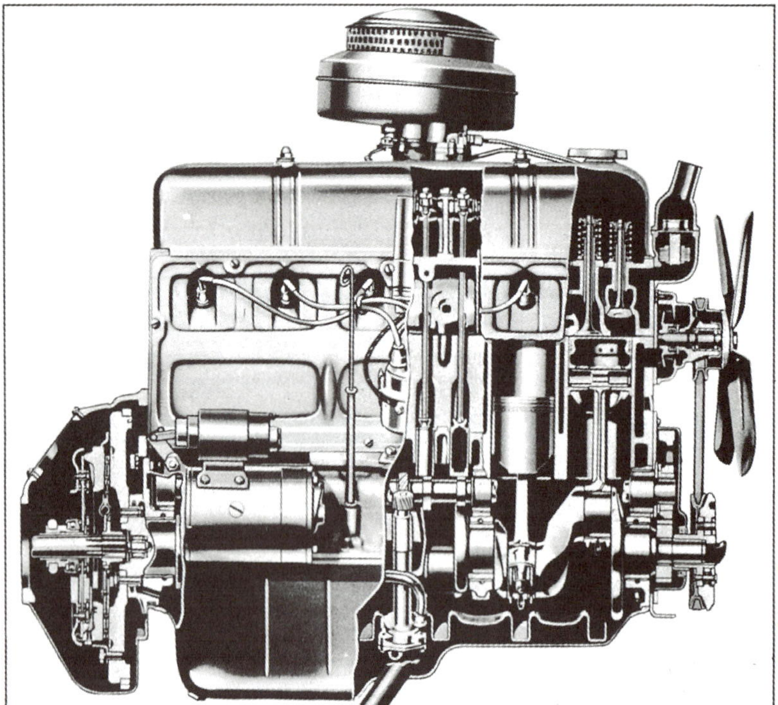

# CHAPTER 2

## THE CHEVY POWER FAMILY

YOU could never look at a little 1928 4-cylinder 35-hp Chevrolet engine and imagine its being worked up to 150 hp. But it has been done. Nor could you easily imagine one of the 3-main-bearing 1934-'36 80-hp sixes being pushed to 200 hp. That *hasn't* been done!

Our story begins in 1937 when a "new deal" at Chevrolet put a rugged 4-main-bearing lower end in their already rugged 6-cylinder block to give a basic power unit that could pump a sustained 200 hp and ask for more. Even then, the block went practically untouched by speed tuners for more than a decade. Only in the last three years has the Chevy suddenly come into its own, to a point where it now rivals the venerable Ford V8 on a HP and lb. basis.

Why was this rugged block forgotten by hot-rodders in their search for light, powerful engines? Certainly it wasn't because parts were more expensive or less-widely available than for the V8. We think the reason lay in the actual *performance* of the Chevy 6 in stock form.

The Chevy was never a "hot" engine from any performance standpoint, as was the pre-war Ford. The Chevrolet engineers, under the direction of the brilliant Alex Taub, designed that engine from a completely detached and practical viewpoint — they wanted a design that would be very cheap to produce, would give a fair peak output, very good lugging at low RPM, and would *stay put* a long time. This meant a "cool" valve timing and heavy pistons and rods. Well, the Chevy engineers got what they wanted — but John Q. Public got anything but a hot stock engine.

As a case in point, you could easily run a pre-war Ford up to 5500 rpm in the gears; put a Chevy up to 4500 and you wonder when it's coming through the hood! What effect do you suppose this had on hot-rodders? They simply steered clear of the Chevy in droves. It took Wayne Horning to prove they were wrong. With light aluminum pistons, he showed that this little six could stay with any V8 — and some say it will even out-wind the latter on the long drag.

So here we are today with a beautiful block, rugged lower end, efficient overhead-valve system, and a fairly light engine weight. It's a "natural" for souping.

### 1937-'40

We're not going to deal in this book with the pre-1937 sixes; these had only three main bearings, long stroke (4 in.), and were generally unsuitable for high outputs. However for 1937, the Chevy engine was completely redesigned with a new head layout, four main bearings, a shortened 3¾-in. stroke, and generally "beefed-up" block. The bore was 3½ in., which gave a total piston displacement of 216.5 cu. in.

With standard compression ratio of 6¼:1, the rated peak output was 85 hp at 3200 rpm for all models. A rocker-operated overhead valve layout was used, as shown in some illustrations. A very mild timing was employed, combined with 30° seats and a lift of 0.31 in.; intake opened 9° BTC and closed 29° ABC, with 52° BBC and 1° ATC for the exhaust.

*Fig. 2-3. Cutaway view of the Chevrolet cooling system; note ample water flow around the valves.*

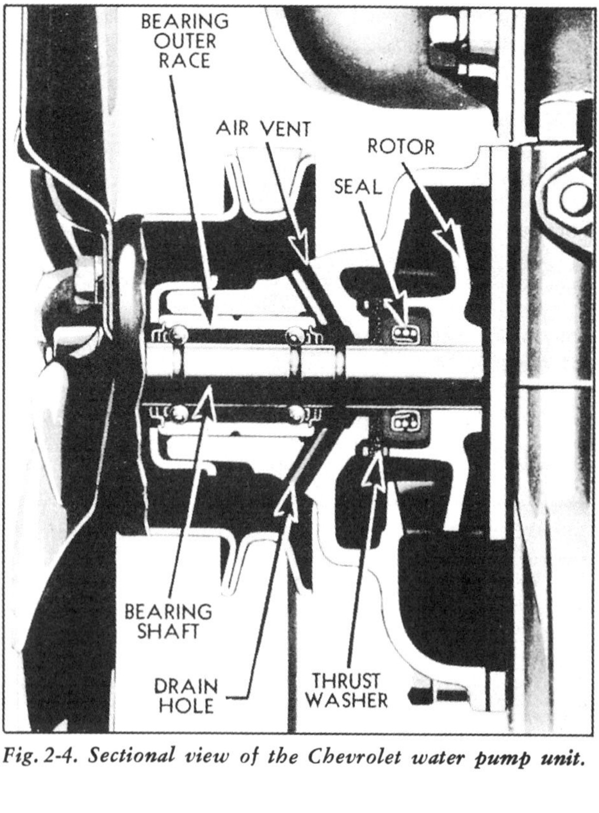


*Fig. 2-4. Sectional view of the Chevrolet water pump unit.*

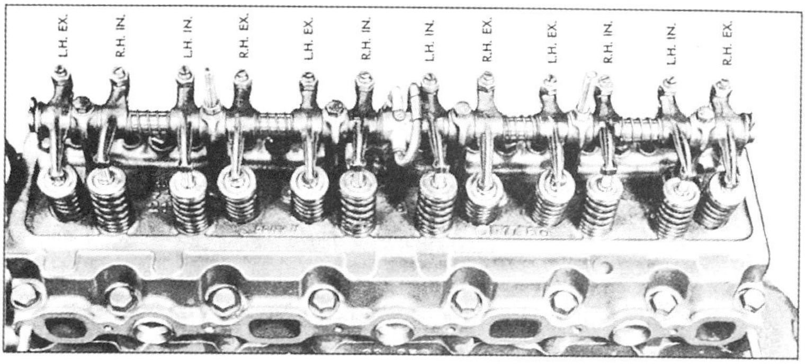


*Fig. 2-5. Head layout of the Chevy showing arrangement of the four different types of rocker arms.*

This compromise gave good torque over a wide speed range, but was poor in the 4000-rpm range. Valve spring pressure at the "open" position was 100 lbs. on the '37 engine, but was raised to 115 lbs. in '38 and to 125 lbs. for 1940. No valve seat inserts were used.

Getting down into the block, we find heavy, 3-ring, cast iron pistons fitted with a 0.002-in. skirt clearance. This setup was beautiful from a wear standpoint, but it killed off the high RPM potentialities. Rods were heavy and rugged too, and were of the cast or "poured" babbitt type, where you replace the entire rod when changing bearings. Bearing areas clear through the engine were huge — in fact, the Chevrolet has more bearing

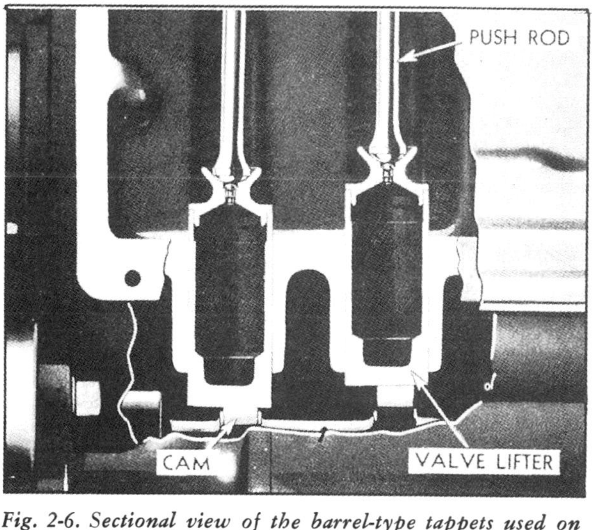


*Fig. 2-6. Sectional view of the barrel-type tappets used on post-1937 Chevy engines (except Powerglide).*

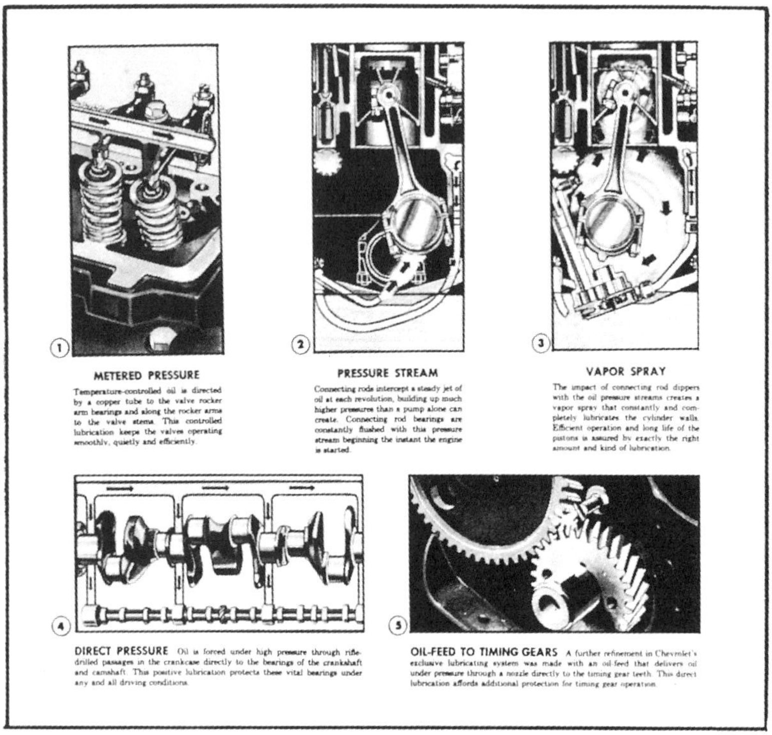


*Fig. 2-7. Features of the novel Chevrolet lubrication system. Don't trust this layout much above 150 hp.*

area per cu. in. of displacement than any other U.S. production engine. The rod big-end was 2.31 in. in diameter and the four main bearings ran from 2.68 to 2.78 in. These main bearings were of the insert type that must be align-bored after assembly.

Lubrication was by a weird but effective combination of splash, jet, and pressure. Jets of oil were aimed at scoops in the rod caps as they came

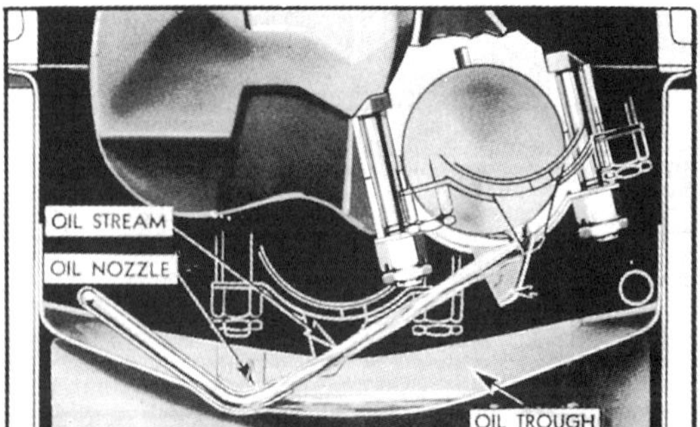


*Fig. 2-8. Detail view showing lubrication of the rod bearings by jet and scoop. The effective feed pressure increases with RPM, but things get hazy over 4000 rpm.*

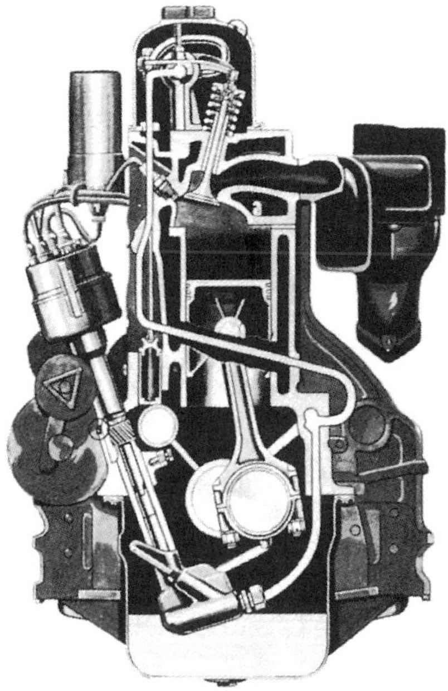

*Fig. 2-9. View showing arrangement of oil lines on the Chevy engine.*

around; then at bottom stroke, the scoops passed through an oil trough. The total feed to the big-end bearing was thus increased to the point where it would handle any lubrication conditions at any RPM on the stock setup (some splash systems are not too reliable at high speeds). For the rest of the lubrication system, a gear type oil pump put a modest 10-15 lbs. pressure to the mains, cam, and rocker arm bearings. It was a very efficient setup, and will even handle a certain degree of souping without trouble.

Ignition was by Delco-Remy to 14-mm. plugs; a combination centrifugal-vacuum spark advance mechanism was used, giving a maximum centrifugal advance of 25° at 3600 rpm. Carburetion was through a standard Carter W-1 single-barrel 1.19-in. model.

1941-'47

For the 1941 model year, to increase compression and combustion smoothness and to decrease carbon deposits, the cylinder head was extensively redesigned. General layout was the same, but the combustion chamber was lowered in conjunction with flat pistons (in place of the former dome-top). This placed the exhaust valve head flush with the piston, with only a gasket thickness separating them. In addition, porting was considerably improved and spark plug position was changed slightly, with small 10-mm plugs being adopted.

These changes in the shape of the combustion chamber raised compression to 6½:1 and boosted the power rating to 90 hp at 3300 rpm. Aside from some changes in the rocker arm, spring retainers, valve guides, etc., necessitated by the head changes, there were no major changes in the rest of the engine.

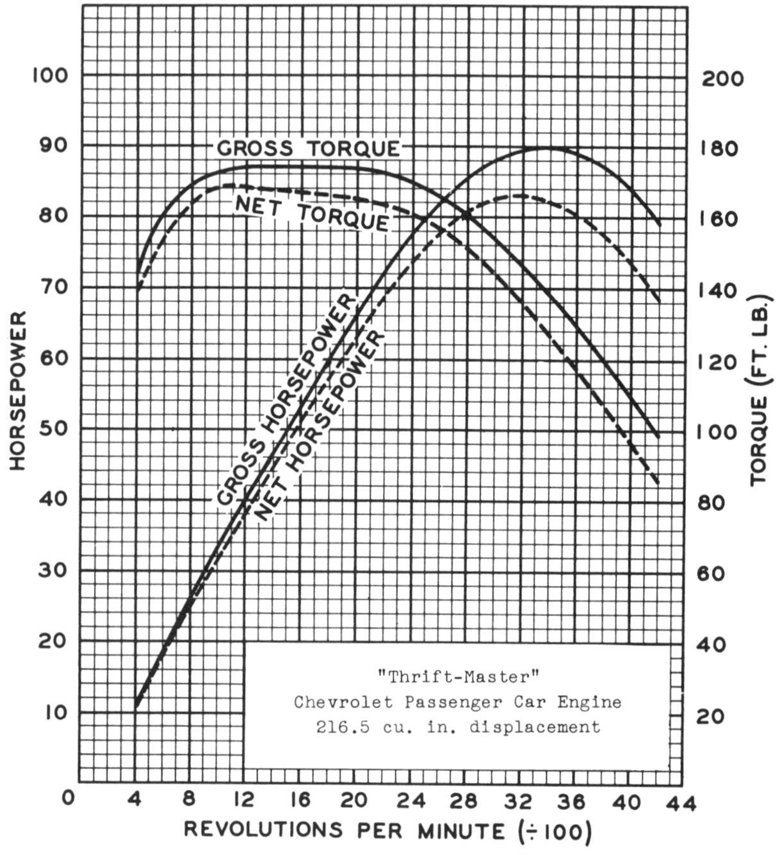


*Fig. 2-10. Officially-notarized performance curves for the post-war 90-hp Chevrolet passenger car engine.*

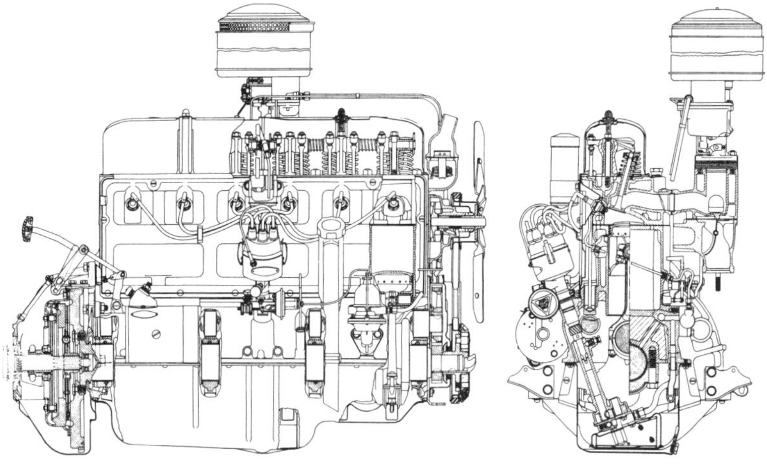

*Fig. 2-11. Sectional views of the 1938 Chevrolet engine. Displacement, 216.5 cubic inches. (Courtesy Automotive Industries.)*

## 1948 AND LATER

One change was made for the '48 models that was significant, and that was the adoption of precision insert-type main bearing shells. This eliminated the need for align-boring and permitted replacing the mains without dropping the crank. Valve timing also was altered a bit, running 1-38 intake and 42-9 exhaust, with 0.29 in. lift. The power ratings were unchanged.

For 1949, some minor head modifications were made, compression was raised a notch to 6.6:1, and the Rochester-GM Model B 1.22-in. carb was adopted – all of which boosted the rating to 92 hp at 3400 rpm. There were no other major changes.

## "POWERGLIDE", 1950 AND LATER

With the introduction of Chevrolet's "Powerglide" torque converter transmission in 1950, they needed an extra shot of brute HP to make up for the slip loss in the fluid converter at low speeds. The obvious answer was the Hi-Torque truck block. This is actually the same basic engine, but has a higher block casting with slightly thicker cylinder walls, 1/16 in. larger bore (3-9/16), and runs a crankshaft with 3/16 in. longer stroke (3-15/16) – which combines to give a total displacement of 235.5 cu. in. On the passenger car version, compression is 6.75:1 and the power rating is 105 hp at 3600 rpm with the 1.34-in. Rochester carb.

(Parts like the crankshaft, rods, camshaft, etc. are interchangeable on the two engines, though the cylinder heads are not; special pistons must be used, however, when you run a Powerglide crank in a small block.) Weight of the Powerglide engine is about 600 lbs.

## WHICH BLOCK?

Just which blocks are preferred for our various souping purposes? Naturally the Powerglide or truck block is preferable for maximum output because of the larger piston displacement obtainable. This is a "must" for racing. Otherwise, there appears to be little to choose between the various small blocks. The '48 and later blocks with the precision insert mains would be slightly cheaper to set up. The 1937-'40 cylinder head design is preferred for reworking, as it has heavier sections for porting, large valves, etc. Since this head can be fitted on later blocks, it makes a pretty good deal for the fellow with a late engine who can't go into special heads.

So now we've had a look at the "raw material" we have to work with. The next job is to find out just what this "power" is, what paths to follow in bleeding more of it out of our engine, and what specific steps we should take — remembering all the time that vital item of *costs.*

# CHAPTER 3

# THE PATHS TO POWER

IF YOU should hear a Chevy off in the distance wailing through twin pipes in a 6000-rpm D#, it might be our Leadfoot Louie trodding his favorite "path to power"! With Louie, RPM is the thing. If nothing else works, just wind it up — the tighter the better. This little theory has cost him more than one block and has never given him a fast car.

Of course, what our moron hero doesn't know about engine performance would fill a book, but we've known a few sharp mechanics who had some pretty wild ideas about power and torque and performance in general. We remember the one who slapped on three carbs to get "more torque below 2000 rpm" — or the boy who wouldn't touch an overdrive because he didn't want the "extra friction caused by increased piston side thrust in the cylinders".

A fundamental knowledge of engine performance would have steered these mechanics straight. Similarly, we've got to start at the bottom if we're going to intelligently determine our various "paths to power" and the specific souping steps we should take.

In the first place, we learned back in Chapter 1, that the one thing we're after in this business is the *maximum possible h.p. at all r.p.m.* That's simple. But to best understand how to get this, we'll have to start at the beginning and trace the process whereby we convert the heat energy in the liquid fuel into a twisting force or torque at the flywheel.

You all know the basic principles here, so we won't go into that; let's take a look inside the cylinder and see how the forces are developed there during the four cycles.

## THE INTAKE STROKE

On the first "cycle," or intake stroke, the piston moves down and sucks in a charge of fuel-air mixture through the carburetor and manifold, under the 14½ lbs./sq. in. atmospheric pressure. Theoretically, then, with wide-open throttle the piston should draw in a volume of gas equal to the cylinder displacement in cu. in., and the final pressure should be about 14½ lbs.

However, things don't happen quite according to theory in the heat and fury of an engine cylinder. This is because the gas mixture possesses weight and momentum, and is subject to turbulence and friction forces in its flow through the manifold and ports. This means an inevitable pressure loss.

To a certain extent, we can take advantage of this fluid momentum to decrease the pressure loss by timing our intake valve to open and close before and after the theoretically-correct dead center points. By opening it 5 or 10° before top center, we get the valve well off its seat by the time the piston starts down, thereby increasing the effective valve area per degree of crank travel; by leaving the valve open 40° or 50° after bottom center, we utilize the inertia or momentum of the incoming gas flow to further fill the cylinder while the piston is coming up.

This valve timing must be a compromise, however, because at low engine speeds, the flow has very little inertia and we'll just pump it back into the manifold by leaving the valve open. Also, our compression can't begin until the valve is sealed. In this case, it would be better to close the valve very near bottom center. So obviously, the valve timing must be carefully selected to fit our performance requirements.

This effect can best be illustrated by considering "volumetric efficiency." If you measure the cubic feet of air entering the carburetor per minute, and then calculate the theoretical volume the engine *should* consume in that time (½ its piston displacement per rev), the resulting ratio or percentage is called the volumetric efficiency of the engine. Obviously it will vary widely at different speeds. Fig. 3-1 shows how the V.E. curves might look on the stock Chevy layout with various intake valve timings.

Now you'll probably ask, "With the cold cam, opening and closing near the center points, it seems like the V.E. should be near 100% at very low speeds. Why is it less than 90% at all times?"

Now you have hit at a big item in all souping work. It's all a matter of gas density. V.E. is measured only on a basis of gas *volume*; therefore when the intake air *expands* in contact with the hot manifold, ports, etc., what you measure as 90 cu. ft. going into the carb may become 100 cu. ft. going into the cylinder! So even though there may be virtually no friction and

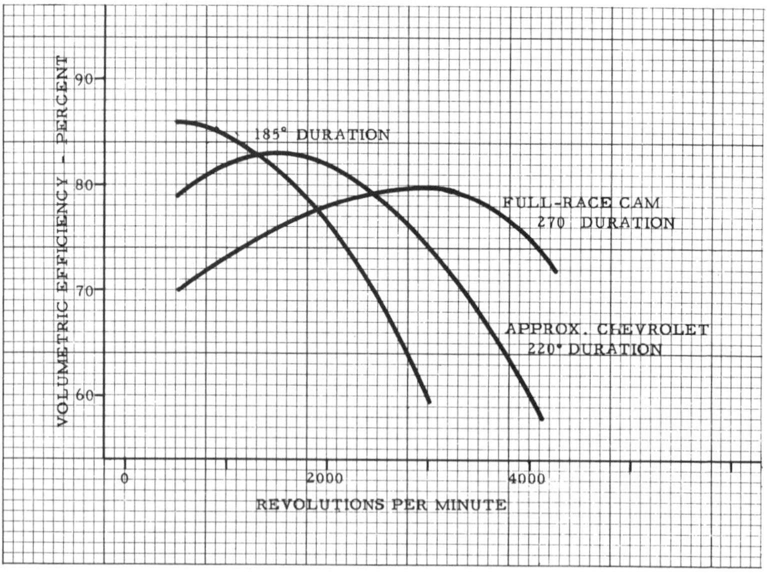


*Fig. 3-1. Typical volumetric efficiency curves (full throttle) for various intake valve timings.*

pressure losses in the air flow at low speeds, the flow does expand from the engine heat so that the true *weight* drawn in is considerably less than the theoretical.

With a given engine at a given RPM, the power output will depend on the weight of fuel-air mixture inducted on the intake stroke. It's vital that you get this concept firmly in mind. It's not the cu. in. you pack into the cylinder — but the *pounds*. This means minimum possible intake air temperatures, and immediately suggests two easy souping steps: (1) Cut off the manifold "hot spot", and (2) lead the carburetor intake stack out of the hot engine compartment.

## THE COMPRESSION STROKE

Getting on to the compression stroke, we find things a little different here. The degree of compression, of course, depends on the "compression ratio" — that is, the ratio of the total cylinder volume at bottom stroke to the total volume at top stroke (which would actually be the combustion chamber). This ratio is about 6½:1 on the Chevy engine, and averages 7:1 for all late U.S. stock engines.

If you remember your high school physics (Boyle's Law), you might suppose our final compression pressure would be equal to the pressure at the beginning of compression multiplied by the C.R. (Or, for example, with a V.E. of 75% and 7:1 compression, our final compression pressure would be: 0.75 x 14½ x 7 = 76 lbs./sq. in.)

However, it's not this way at all. Compression generates heat; if the compression is done quickly, as it is in a cylinder, this heat doesn't have much time to leak away, and is largely retained in the mixture so that the heat energy causes the total pressure to increase much faster than slow compression would. It follows approximately the following law:

$$Pc = Pi \text{ (C.R. 1.35)}$$

Thus with 75% V.E. at 7:1 C.R., the final compression pressure would be, not 76 lbs./sq in., but nearly 150 lbs. So compression pressure rises very quickly as you raise C.R.; boosting the ratio 20% will raise the pressure 28%. This is a very handy souping tool.

## THE POWER STROKE

On the "power" or expansion stroke, we ignite our compressed mixture and convert the internal heat energy of the fuel into pressure on the piston head. However, since it takes a split second for the charge to burn through, and since we must have the combustion *completed* as near as possible to top center for maximum power and fuel economy, we must advance the ignition point from 5° to 40° before top center, depending on speed and load. (The ideal situation, of course, would be to have instantaneous combustion at top center.) In this way, combustion is virtually completed within 10° after top center, and the rest of the stroke is merely expansion of the gas from high pressure — and that's where we get the torque on the flywheel.

The peak pressures reached in the cylinder during combustion are generally three to five times the compression pressure, or about 600 lbs. in a stock engine. On the expansion stroke, this pressure drops off approximately by the same exponential law as for the expansion stroke.

## THE EXHAUST STROKE

The exhaust stroke calls for no particular comment other than a consideration of valve timing. In this case, we must open the cylinder and relieve the pressure *before* the piston starts up on the exhaust stroke — otherwise it will be starting up against a pressure of some 45 lbs./sq. in., or a total of 430 lbs. for a 3½-in. piston. By opening the exhaust valve around 40° BBC, we reduce the cylinder pressure to near atmospheric on the first part of the up-stroke, and increase the power slightly. We also leave it open a few degrees ATC to utilize the inertia of the gas column to further scavenge the cylinder.

Fig. 3-2 shows some typical curves of cylinder pressure on a stock engine; notice that increasing the total area bounded by the banana-shaped curves, such as increasing C.R., will boost the HP.

## CONVERTING CYLINDER PRESSURE INTO TORQUE

Now that we've seen how that heat energy in the fuel is converted into pressure energy on the piston heads, it remains to convert this linear force into a rotational force or torque on the flywheel. As you know, this is accomplished through a piston-rod-crank train. Sounds easy — but there's

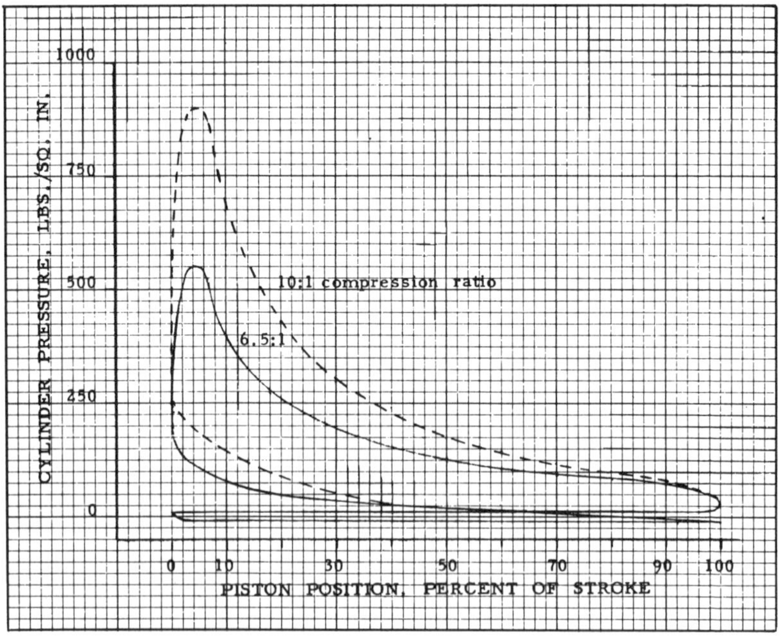


*Fig. 3-2. Curves showing how cylinder pressure varies during the four cycles in a typical stock engine; full throttle at 3000 rpm.*

much "frictional grief" involved in this business before we finally get the desired torque.

These total losses are composed of two distinct factors: (1) Old-fashioned friction between piston and cylinder walls, in bearings, between cam and tappets, in cam and distributor drives, gears, etc., and (2) "pumping" losses caused by the pistons moving against a vacuum on the intake stroke and a pressure on the exhaust stroke.

The friction losses do not vary appreciably as the load on the engine is changed. However, our total mechanical losses vary widely with engine speed — being approximately proportional to the *square* of RPM. In other words, if a mere 10 hp is lost in friction at 1500 rpm, we'll be losing some 90 hp at 4500 rpm. There is not a lot we can do about this in souping, but these figures should at least serve to focus attention on this important item, and to urge every possible step to combat friction, a deadly enemy.

Also in this direction, a couple of fuel economy moves suggest themselves: (1) Run the engine at the lowest possible RPM on the road (overdrive) so we won't be burning a lot of gas just to overcome useless friction, and (2) use a dual straight-through exhaust system to reduce "back-pressure" on the piston heads — a move that should increase road HP and fuel mileage by some 10%.

So now we've traced the power problem from the energy in the fuel to torque on the flywheel. Now let's take a closer look at these concepts of power and torque with an eye to setting up our paths to *more* power:

## POWER AND TORQUE

Power is the *time rate of doing work.* If someone built a chicken coop in six hours, we would have a hard time estimating his HP output on the slide rule — and probably our only comment would be in regard to the *quality* of a rush construction job! But if he said he could lift a 60-lb. crankshaft up 3 ft. in one second, we would quickly quote his rating as about 1/3 hp.

Work is a force exerted through a certain distance, and is measured in *foot-pounds.* The U. S. *horsepower* is officially defined as 550 ft.-lbs. of work per second. In other words, in lifting the crankshaft, you have done 60 x 3 = 180 ft.-lbs. of work per second, or about 1/3 hp. Or similarly, if a hot rod is tearing along at 90 mph (132 ft./sec.) and the total drag on it at that speed is 350 lbs., then the engine is doing 350 x 132 = 46,200 ft.-lbs. of work per sec., or 84 hp.

Torque is quite another thing. This is a force exerted about a center of rotation, and is measured in *pound-feet* (the opposite of the term for work). For example, if you exert a push of 30 lbs. on the end of a crank arm 1½ ft. long, you would be exerting 30 x 1½ = 45 lb.-ft. of torque.

A torque force must move through a certain arc distance, say one revolution (6.28 ft.), before it becomes work — and how fast it moves through this distance determines how much *power* is being produced. (You must understand that torque can be exerted without any motion at all, but the power output is *zero* in this case; for instance, a gas turbine engine exerts its maximum torque with the power turbine *stalled.*)

*Fig. 3-3. Taylor water-brake dynamometer; note the scale for measuring torque and the valves for adjusting the load.*

*Fig. 3-4. Full water-brake engine test setup.*

Torque is related to the U. S. horsepower by the following formula:

$$HP = \frac{T \times RPM}{5250}$$

For example, with 170 lb.-ft. of torque and 3500 rpm, the HP is: HP = 170 x 3500/5250 = 113½ hp. If you check any of the power and torque curves in this book you'll see that they are related by this formula.

## PERFORMANCE TESTING

How are these technical concepts related to what the engine actually will put to the flywheel? Let's put it on the "dynamometer" and see.

A dynamometer is merely a machine for absorbing and measuring torque. The resistance or torque-absorbing unit is mounted in a cradle and is free to rotate, being restrained only by a spring scale. The engine being tested is coupled directly into the resistance unit. This rotational resistance can be set up in a number of ways; by generating electric current and dissipating it as heat in a resistance grid, by an electro-magnet, by churning water in a turbine, or even by a brake band on a drum. All these methods merely absorb the power output and get rid of it in the form of heat.

With the resistance unit absorbing the output of the engine, the torque *reaction* will tend to turn it in its cradle. This torque reaction is restrained and measured on the spring scale, and the HP output is then calculated from the above formula. Figs. 3-3 and 4 show typical "water brake" dynamometer installations with their measuring scales.

In running a power curve test, the engine is operated at full throttle. The resistance of the dynamometer is adjusted to hold the engine at a certain speed, perhaps 500 rpm, and the torque is read; this is repeated at speed increments of possibly 500 rpm (1000, 1500, 2000, etc.) till the full torque curve is obtained. The power curve is calculated from this and the results plotted on a graph. Fig. 3-5 shows what we might get with a typical stock setup. Notice carefully the *shape* of the curves – this is vital in souping work.

If the torque is dropping, how can the power be going up? This is all a matter of *rates*. Since HP is proportional to both torque and RPM, then if the torque has dropped only about 10% while the RPM has increased 20%, the power will be going up. When the two rates of change are the same, the power is at the "peak".

Since torque is the basic performance factor, let's study the curve more closely:

## THE TORQUE CURVE

Notice first the shape of the curve – how it reaches a peak at about ½ the peak power RPM, and that this peak is something around 25-35% above the torque at peak HP. Why is it shaped this way? To answer this, we must put together some of the facts about cycle pressures we learned earlier.

The torque curve drops off at high RPM for two major reasons: (1) Decreasing volumetric efficiency, and (2) increasing friction. The first factor lowers the compression pressure, thereby lowering the effective

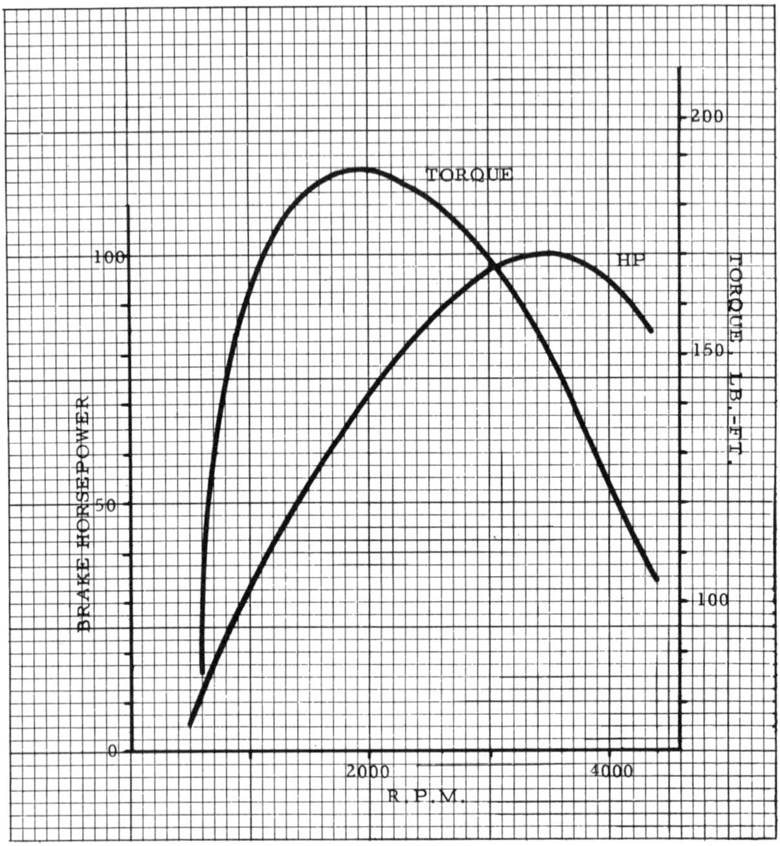


*Fig. 3-5. Typical dynamometer test curves for a stock car engine.*

cylinder pressures on the expansion stroke. Friction, of course, eats up an increasing proportion of the total power generated in the cylinders as the RPM goes up. The net result is a swift drop in torque at the flywheel above about 2500 rpm.

The drop in torque at low RPM is a different matter. Here friction has practically no effect one way or the other. The major gremlin instead, is valve timing. In order to get a decent peak HP per lb. of engine weight, and to get good gas mileage (things we must have in a stock car), we should overlap the intake and exhaust valve opening and leave the intake valve open 30-50° ABC. This is fine at 3000-4000 rpm, but it is rough at 1000! The incoming gas flow has little inertia and just backs out of the valve when the piston comes up—and since compression can't begin until the valve is seated, our effective compression ratio would be reduced from around 7:1 to 6.2:1. We can't have everything, even in engine performance!

A secondary cause of the low-speed drop of the torque curve is a loss of compression pressure due to the compression heat having more time to leak away through the cylinder walls. This is minor, but you'll find that engines with large bores—that is, having less sq. in. of cylinder surface area per cu. in. of volume—will give better low-speed torque than long-stroke engines, other factors being equal.

Now we have seen how the fuel energy is transformed into torque, and how it gets to the flywheel.

## THE PATHS TO POWER

Let us once again stress this matter of gas *density.* It's an important concept in souping. Density refers to the weight per unit volume of a substance—or say lbs. per cu. ft. It means nothing to say that an engine burns 250 cu. ft. of fuel-air mixture per minute. That volume of mixture would weigh 12 lbs. on Pikes Peak on a warm day, and 22 lbs. on a cool day at sea-level! Quite a difference. It's the weight we burn per unit of time that determines HP. Look again at the HP formula:

$$HP = \frac{T \times RPM}{5250}$$

Let's break it down a little. Obviously, torque will depend on three factors: (1) The weight of charge inducted on each intake stroke, (2) the efficiency with which the charge is burned in the cylinder, and (3) the mechanical efficiency or friction in the engine. Also, that first factor, charge weight, will depend not only on the volumetric efficiency and charge density, but on the piston displacement. In other words, larger cylinders will draw in a greater charge weight, given constant V.E. and gas temperature.

From here it's a very simple matter to set up the five basic paths to power: (1) Increase piston displacement, (2) increase the weight of fuel-air mixture inducted on the intake stroke, (3) increase the efficiency of combustion, (4) increase the RPM potentialities, and (5) decrease friction and pumping losses.

The next thing is to break these five paths down into individual steps that we can use in souping a stock Chevy block. These steps fall right into place:

FOR INCREASING PISTON DISPLACEMENT:

1. Boring and stroking

FOR INCREASING WEIGHT OF CHARGE INDUCTED:

1. Draw cool air into the carbs, and use no manifold "hot spot"
2. Burn fuel with high latent heat, such as alcohol
3. Enrich fuel mixture, within limits
4. Increase total carb throat area
5. Enlarge flow path area of manifolds and ports
6. Smooth manifold and port walls and have easy radii in the flow path
7. Reduce valve stem and guide obstructions in the port
8. Increase discharge area of the valves
9. Alter valve timing and rates according to performance requirements (no valve floating)
10. Supercharging

FOR INCREASING THE EFFICIENCY OF COMBUSTION:

1. Raise compression ratio
2. Proper spark timing and sufficient spark voltage
3. Proper air-fuel mixture ratio (about 12:1 for gasoline)
4. Burn fuel which will allow no "knock" or preignition
5. Good exhaust scavenging

FOR INCREASING R.P.M. POTENTIALITIES:

1. Lighter pistons and rods
2. Perfect static and dynamic balance in crank train

FOR DECREASING FRICTION AND PUMPING LOSSES:

1. Adequate lubrication at all times (which may mean 60-80 lbs. oil pressure to all points)
2. Use lighter oil, within limits
3. Increase bearing clearances
4. Eliminate exhaust back-pressure

Right at this point, we might also mention the steps for increasing the *acceleration* potentialities of an engine—which is also an important consideration in building up any hot vehicle. All the above steps are concerned only with increasing power output at *constant* RPM. When accelerating, however, we have to consider the *weight* of rotating and reciprocating parts, and reduce these weights to the very minimum. These parts include the flywheel, crankshaft, rods, and pistons. There's not much we can do with the latter parts, but we can "chop" that flywheel to half its stock weight without too much trouble.

Now before we get into the real souping business, we should probably mention the three rigid limitations on HP that will be faced with any road engine: (1) Fuel, (2) mufflers, and (3) cam timing.

FUEL—With a road engine, we will have to feed on pump gasoline available at any filling station—which means a maximum of about 82-octane. This will limit compression ratio if we are to avoid destructive knock or pre-ignition on full throttle.

*Fig. 3-6. Two views of the full-race Wayne conversion of the Chevy block. Note triple carburetion, dual exhaust, and dual-coil ignition.*

MUFFLER – Most states have strict laws about exhaust silencing. Stock exhaust systems are rough on HP at high speeds, but the dual straight-through replacement systems on the market virtually eliminate back-pressure and will still barely satisfy the law. They are a hefty investment in some cases, but a "must" for a highly-souped engine.

CAM TIMING – Cam timing must be a compromise between good power output at high speeds and low speeds. We can't design a cam to be at its best at both "ends." If we lengthen out the valve open periods to boost peak HP, the torque at low RPM suffers. In fact, we can carry this cam timing business to a point where the peak torque will be *above* 3500 rpm – and it will barely idle at 1000. This would mean "idling" at about 22 mph in high gear. To top it all, it's mighty easy to "kill" an engine like this in city traffic, even if you ride the gear shift hard. In other words, cam timing for a road engine is limited by the need for some torque at low RPM.

We know now just where we're going and what steps we must follow. The next thing is to dig right into that block.

*A serious contender on Southern California dirt tracks is the "Cannon-Ball Special", equipped with 270 cu. in. Wayne-converted GMC power plant.*

*Built only for drag meets and dry-lake runs, this Model T-bodied hot rod has full-race 270 cu. in. Wayne-GMC engine in rear, has made nearly 150 Mph! Called "The Beast".*

*A track roadster of consistent performance, No. 44 has 270 cu. in. GMC engine.*

# CHAPTER 4

## FITTING UP THE BLOCK

LEADFOOT LOUIE was always a dyed-in-the-wool Ford V8 man until a couple of months ago. He liked that short, sturdy crankshaft on the "V" block layout. But he always got a bad case of piston slap and oil consumption after 5,000 miles with his 0.010-clearance, 3-ring racing pistons. Lately, he figured out why: The weight of the pistons when they slide up and down in the cylinders causes them to wear faster on the bottom side! Now you can't get him near a V8 block.

But seriously, Louie could have done a lot worse than to turn to the Chevrolet block. Here's a rugged lower end with huge bearing area and a stiff, short-stroke crank that will really *stay put*. It's proving more conclusively every day that it is ideal raw material for the souper.

In this chapter, we want to go right through that block from top to sump and see what we can do with it for more HP. Just to keep the record straight, this will probably be the most important phase of the whole souping job. It doesn't take a genius to fire up 10,000 lbs. on the piston head, but it takes more than wishing to put that kind of force through the crank train to the flywheel, and still keep things together. In other words, a hot engine is no hotter than its block setup. So let's pay strict attention to this vital business—it will pay off in durability and reliability as well as HP.

### SELECTING A BLOCK

If you are just souping up the family bus a bit, you won't have much of a choice in your block selection. The block is there in your Chevy—soup it or leave it. Even the boy with limited funds who's stripping his car down to the "hot rod" category may not be able to afford a brand new block to build on. For you fellows, a word about the condition of this block you're souping:

When you increase the output of an engine, you increase the normal operating forces clear through that engine—which means that we can't neglect the clearances and expect it to stay together. Where you might get by with cylinder bores 0.010-0.015 in. tapered at the top with the stock setup, we wouldn't want anything over 0.004 in. out when we soup.

So if you are tearing your engine down for the souping job, check the dimensions and act as follows: If the bores are more than 0.002 in. out-of-round at any point, better rebore. If rod or main bearing journals on the crankshaft are worn over 0.001 in. on taper or out-of-round, have them reground to the next standard undersize. Rod and main bearings themselves should be replaced if more than 0.002 in. out. If you're doing a very mild souping job and won't have occasion to tear the engine down, we don't suggest your ripping it apart just to check clearances unless you are in great doubt; if the engine has less than 30,000 miles on it and the oil consumption and pressure are not out of line, we wouldn't worry about it.

If you're building up a hot road or track car from scratch, you'll probably want to start with a new block. Theoretically, any old block will do just as well, assuming dimensions are all "on the money". But machinery gets tired. Fatigue and corrosion ruptures set in and before you know it, you have a cracked block, broken crank or rod for no decent reason. In other words, all the reboring, refacing, regrinding, and refitting in the world won't make a sure-bet world-beater out of a 10-year-old block with 150,000 hard road miles on it! If you're doing a big souping job, we definitely suggest a new block (including crank and rods).

Otherwise, there seems to be little choice between one block and another except this: Check up in the block casting where the cylinder skirts extend down into the crankcase on the camshaft side; the best block for souping is that in which all the bores come as near as possible to the exact center of their casting cylinders. (The bore will often come quite far off due to core core shift or "creep" in the foundry.)

## BORING

Back in Chapter 3 we learned that increasing piston displacement is a basic souping step, since it increases the weight of fuel-air mixture inducted on the intake stroke. The displacement formula for a 6-cylinder block is:

$$D = 4.72 \times S \times B^2 \text{ (cu. in.)}$$

where "S" is the stroke in inches and "B" is the bore.

Obviously, total displacement will increase more rapidly with increases in bore than with equal increases in stroke, since displacement is proportional to the square of the bore. For instance, on the standard Chevy block, a ⅛-in. increase in bore will boost displacement 7%, while the same stroke increase would raise it only 3%. This illustrates the preference for bore increases over stroke increases; inch for inch, boring is a much more effective and economical souping method.

From the standpoint of engine performance, there is one thing about increasing displacement that we should get firmly in mind: *It is one of the few souping steps that will increase the torque at all R.P.M.* Methods like multiple carburetors and special cam timing boost output at high speeds, but do nothing for the output at low RPM. Increased displacement, on the other hand, by actually burning a greater charge weight per revolution, increases HP over the full RPM range. Fig. 4-1 illustrates this effect, comparing the over-all power curves obtained when you boost peak HP 15% by increased displacement or with altered cam timing. This fact is good to remember if you're building up a road engine for city driving where you need plenty of torque at 1000 rpm.

How about maximum bore recommendations? We must be careful here not to get the cylinder walls too thin in any spot to prevent cracking them at high speed (thin cylinder walls will actually start to whip and chatter under high-output conditions). Casting inaccuracies at the foundry often bring the factory bores slightly off center in the cylinder casting so that the wall thickness will vary. A certain overbore might work fine on one block — but might crack a cylinder wall the first time up to 5000 rpm on another.

So we must limit our maximum bore to a safe figure in all cases. This is usually considered 5/32 in. (0.156 in.) on the standard block and ⅛ in. (0.125) on the Powerglide or Hi-Torque engine. Several companies supply special pistons up to this oversize, adaptable to both standard and

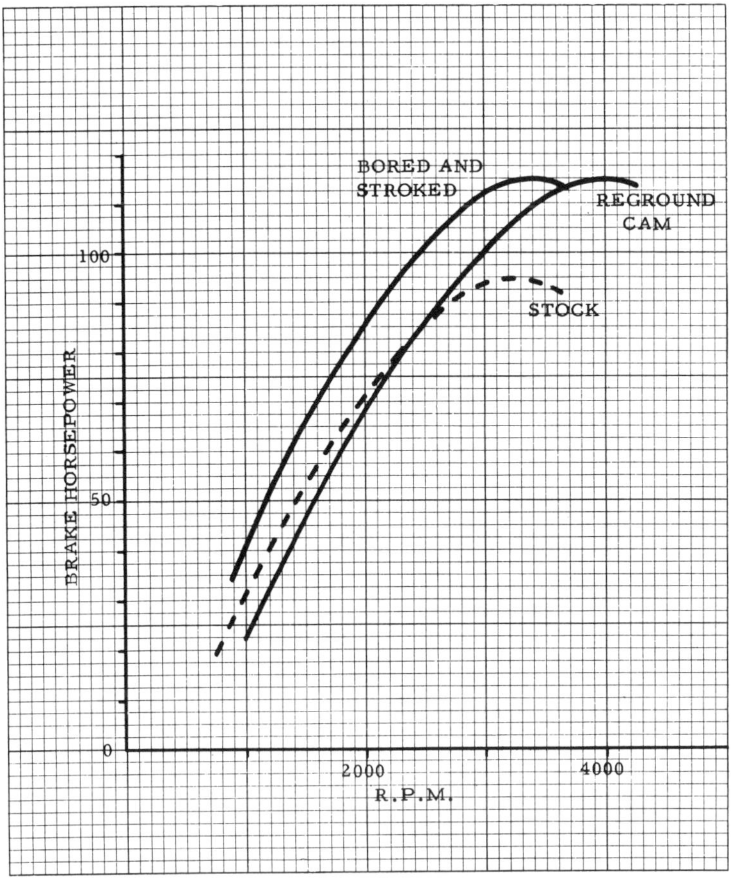


*Fig. 4-1 What happens to low-RPM output when you raise HP by 15% with a reground cam, or by increasing piston displacement.*

Powerglide crank strokes. For smaller overbores, Chevrolet supplies stock replacement pistons up to 0.040 in. oversize. The charge for boring a Chevy block is usually $12-$15.

(Incidentally, some have bored the cylinder wall right out and fitted ⅛-in. wet sleeves for a ¼-in. overbore and a total displacement of up to 270 cu. in. with the Powerglide crank. This is a terrific and expensive job of questionable practical value, so we don't recommend going into it except for the most extreme reasons.)

## STROKING

Here's a great trick with the Ford V8 that is no good on the Chevy. As you know, "stroking" refers to grinding the crankpins down off-center, away from the crank axis, giving a stroke increase of twice the distance the center is displaced. However, the crankpin diameter is reduced by about the same amount, which means a different rod bearing size. On the late Fords, you can stroke ⅛ in. and use stock pre-war rods, but no such possibility presents itself with the Chevy. Besides, with the Powerglide crankshaft, you couldn't increase the stroke more than a very small amount without having the rod shank contact the cylinder base at high RPM.

The Chevy *has* been stroked by building up the crankpins with a "metal spray" or weld and relieving the cylinder wall, but it's a complicated job and hardly worth the bother under ordinary circumstances. So stroking is out of the souping picture on the Chevy block.

Here is a table of piston displacements with various bores and strokes:

TABLE I

CHEVROLET PISTON DISPLACEMENT (CU. IN.)

| | With standard crank; 3¾" | With Powerglide crank; 3-15/16" |
|---|---|---|
| Standard block: | | |
| Stock (3½ in. bore) | 216 | 227 |
| 1/16 overbore | 225 | 235 |
| 3/32 " | 228 | 239 |
| 1/8 " | 232 | 244 |
| 1/4 " (wet sleeves) | 248 | 260 |
| Powerglide block: | | |
| Stock (3-9/16 bore) | ... | 235 |
| 1/16 overbore | ... | 244 |
| 3/32 " | ... | 248 |
| 1/8 " | ... | 252 |
| 1/4 " (wet sleeves) | ... | 270 |

## THE CRANKSHAFT

One major virtue of the Chevrolet lower end from a souping standpoint is the short stroke and large bearing diameter, which give overlapping crank journals and a very stiff, rugged shaft.

We can't help but remember some experiences with the 6-throw crankshaft for the Allison V-12 aircraft engine during the last war. To keep engine weight to a minimum, the shaft was built extremely light — in fact, not much heavier than your Chevy shaft. When outputs began to get up around 1,500 hp at 3000 rpm in the early '40s, this little shaft took to whipping every which way like so much rubber — and main bearings burned up right and left. Rather than beef up the shaft, they *rounded the main bearing shells* in a lengthwise direction and solved the problem in one fell swoop!

We don't face anything like this with our Chevy, but it helps us to appreciate a good, stiff crankshaft.

For souping purposes, the Chevy crankshaft is generally used in stock form without any changes, and has been found to give completely satisfactory service. (For the heavier souping jobs, we'll be modifying it for lubrication purposes only.) The standard 3¾ in. and 3-15/16 in. Powerglide cranks are completely interchangeable in both blocks, though you must use their corresponding camshafts in conjunction (you get a light contact using a standard cam with the Powerglide crank).

Also, you couldn't use standard pistons in conjunction with a Powerglide crank in a *standard* block, since the longer stroke would bring the piston in contact with the head. Several companies supply special pistons for various bores and for standard and Powerglide stroke, so there would be no parts problem here. You can buy a Powerglide crank for about $37; if you are tearing your engine down and plan to rebore, spending the extra for 3/16 in. longer stroke is a practical idea, since you'll have to have special pistons anyway.

Otherwise, there appears to be little we can do with the shaft itself to improve its performance. It might be possible to lighten it 5 lbs. or so by grinding metal off the heavy sections of the crank arms, to improve acceleration. This is not widely practiced and we won't recommend it one way or the other, since the performance improvement would be very slight at best. (If you do this, however, remember to have the crank rebalanced.)

We think the best idea is to use the stock Chevy crank *as is,* except for possible lubrication modifications.

## CONNECTING RODS

The stock Chevy con rods themselves have been found to work very well in souped engines, and there is very little we can do to improve them. Their shank strength seems to be adequate for a sustained output of 150 hp at 4000 rpm, assuming the rods are fairly new and in perfect alignment; we'd suggest new rods for your souped engine if possible. (For an analysis of the rod bearings, see the section on "bearings.") Incidentally, when using the stock Chevy rod with locked piston pin, the pin fit in the piston boss should be a palm-push fit at room temperature. This is vitally im-

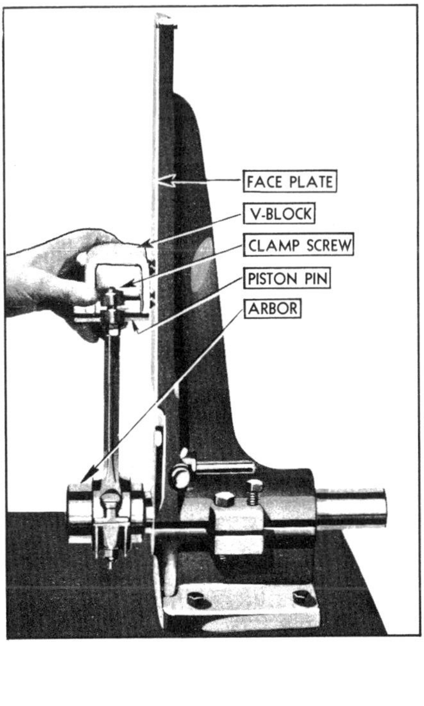


*Fig. 4-2. Setup for checking rod alignment with Chevrolet shop fixture.*

*Fig. 4-3. Wayne-degreed harmonic balancer replacement unit; this is handy on competition engines for checking valve timing, spark advance, etc.*

portant, since there can be no movement at the rod; too tight a fit in the piston could wreck an engine when it heats up.

For outputs of well over 150 hp and for sustained speeds of around 5000 rpm, the stock Chevy rods are not going to be too satisfactory. There are several reasons: The piston pin is locked in the rod and does not float; the rod shank is not of exceptionally stiff section; and the thick "poured" bearings are not suitable for high outputs.

Fortunately, there is another stock connecting rod right at our fingertips that will fill requirements to a "T", and that we can use nicely in the Chevy block. Rods for the GMC Models 228, 248, 270, and 302 truck engines can be adapted with some modifications. These are rugged forged "H"-section rods, rifle-drilled for pressure lubrication to the floating pin, and running precision insert bearings on the stock Chevy crankpin diameter (2.31 in.). And we get all these advantages within a weight increase of only four ounces over the stock Chevy's 31 oz.!

Rod center-to-center length is 7 in. here, compared with 6.81 in. for the Chevy, and pin diameter is 0.990 in. compared with 0.865, so we can only use the GMC rods in conjunction with special Venolia pistons (and the stock GMC 248 piston pins). Also we must drill the crankshaft for pressure lubrication. The rod bearing width is 1/16 in. less than on the stock Chev rod, but this extra side clearance has not caused any trouble on even the hottest 250-hp jobs. These GMC rods run about $5 each, and we would definitely recommend them for track engines of over 200 hp.

A word about rod "polishing:" The idea here is to eliminate the slight friction torque required to accelerate oil that clings to the whirling rods; by smoothing and polishing the rod surfaces, this friction is decreased. We have no dynamometer figures at hand to prove this effect on HP, but the idea is logical. (But don't expect to knock a fifth of a second off your lap time or increase your top speed 8 mph by just ripping out the rods and buffing them off a little!)

## PISTONS AND RINGS

Here's where we can put all kinds of extra pep in our lowly Chevy block with one blow. The relatively heavy rods and cast iron pistons used in the Chevy have always hindered it in the RPM department. In fact, we think the lack of available aluminum replacement pistons was one big reason why soupers shunned this fine engine for so long. We don't like to compare, but we can't help but emphasize this matter of piston and rod weight by measuring the Chevy against the Plymouth, which has just about equal piston displacement. Total piston-rod weight is 59 oz. on the Chevy, compared with 47 on the Plymouth using an aluminum piston; this means a difference of about 300 rpm.

Aside from this matter of RPM, however, there are several other reasons for choosing between stock cast iron pistons and special aluminum jobs for our souped Chevy engine. The iron pistons are fitted with close 0.002 in. skirt clearance, which means somewhat quieter operation and less oil consumption; also the iron will wear longer than aluminum. None of these items, however, will be too important unless you are after a very economical road engine.

Aluminum pistons, on the other hand, are fitted generally with 0.005-0.010 in. skirt clearance; noise and oil consumption go up a bit and wear is considerably faster, especially with large clearances. (You can run split-skirt aluminum pistons with 0.002 in. clearance, but friction is higher than with iron.) The main advantage of aluminum though, is that the pistons weigh only about 18 oz. (compared with 28 for stock), giving higher RPM and slightly better acceleration. And since aluminum conducts heat 4.4 times faster than iron, we get a cooler combustion chamber and reduce the "knock" tendency somewhat.

So that's the score-sheet—make your choice. Personally, we'd recommend the aluminum type if you can possibly afford it. As of now, a number of companies are supplying special aluminum pistons for souped Chevys. Prices range generally from $25 to $42 a set, and they are available in oversizes up to 5/32 in. for use with both standard or Powerglide stroke. One type is designed primarily for road use with split skirt or "T"-slot

*Fig. 4-4. Typical split-skirt aluminum rod piston for standard, overbores, and overstrokes on the Chevy.*

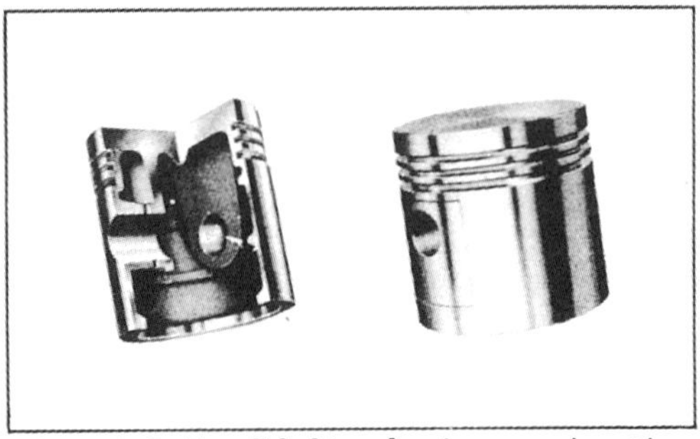

*Fig. 4-5. J. E. solid-skirt aluminum racing piston for the Chevy 6.*

(see Fig. 4-4), and are usually fitted with 0.005-0.007 in. skirt clearance. (A very recent road type incorporates a specially-shaped dome that extends into the combustion chamber to raise compression ratio to 8.2:1; we can't offer reliable data on these as yet.) The track type (see Fig. 4-5) have split skirts and run 0.008-0.010 in. clearance.

The piston ring setup with these special aluminum 3-ring pistons generally follows stock practice—that is, two ⅛-in. compression rings and a 3/16-in. oil control. For racing, pistons are usually designed for narrower compression and wider oil rings; 1/16 or 3/32 in. compression and 5/32 oil rings. Ring gap is 0.010-0.020 in. Follow the ring manufacturer's instructions closely on setup, fits, and clearances.

## LUBRICATION

They say oil is the "life blood" of an engine—and you must get the right amount of the right kind to the right places at the right time, or no engine is going to stay together long at 5000 rpm. Anyone who looks at the stock Chevrolet lubrication system and then thinks of 5000 rpm is very apt to throw up his hands in horror. The Chevy system doesn't *look* good. But it seems to *work*. In fact, this layout has been found entirely adequate for moderate road outputs to over 150 hp.

This system consists of 10-15 lbs. pressure to the mains, a throttled pressure to the rockers, and a jet-scoop pickup for the rods. In the past, the "splash" system for lubricating rod bearings has not been too dependable when the engine was souped up somewhat, principally because the high RPM whipped the oil to a mist in the troughs. With Chevrolet's unique jet system, however, we get a more or less constant and full feed to the rod bearings under all conditions (although even here, things begin to get hazy around 4000-5000 rpm!).

Of course, the success of the stock Chevy lubrication system depends on the proper adjustment of the "aim" of the jets, and this should be carefully checked during rebuilding procedures. (See Fig. 4-6). Other than this precaution, the stock system should be adequate for most road applications.

As we move up into the 200-hp bracket with track engines, however,

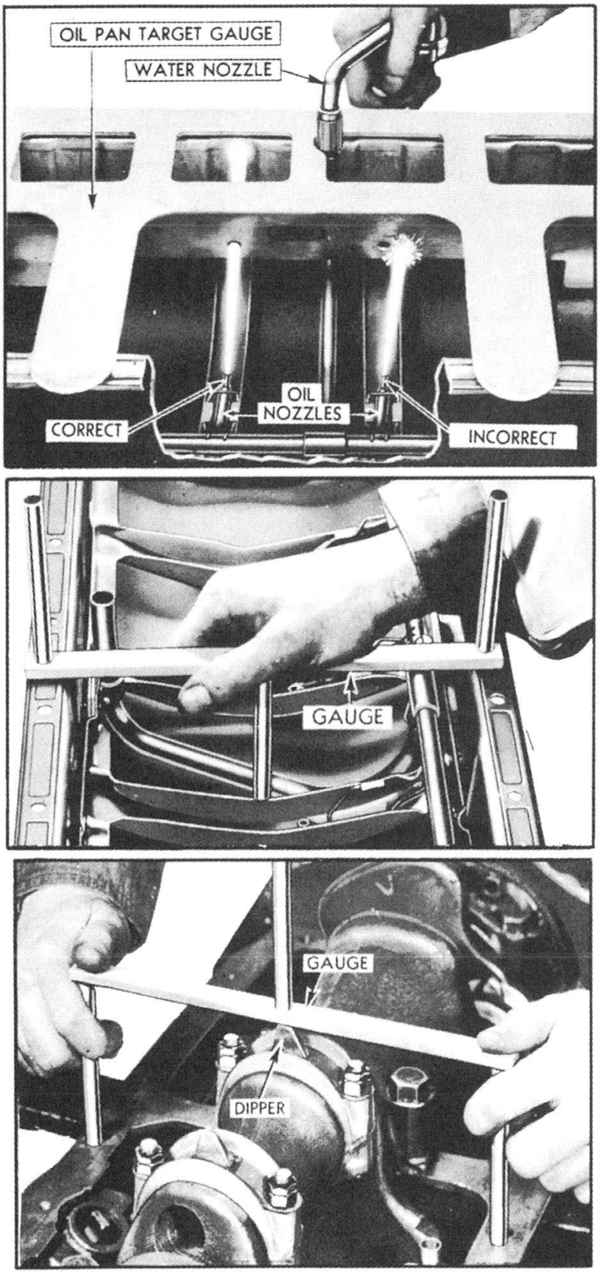


*Fig. 4-6. Method of checking setting of oil scoops and jets in the Chevy rod lubrication system, using special shop fixtures. Be careful about this job — it's got to be right.*

the "jet-splash-pressure" system will not be satisfactory. We will need a lot more feed pressure and oil flow rate, not only to carry the terrific bearing loads, but to help cool the lower end. There is nothing we can do with the stock Chevy lubrication *as is* to fill the bill, so we must convert the whole system to "pressure" (which includes pressure to the rods). Actually, this is not as big a job as it looks, and there are several ways of proceeding:

In the first place, to get the oil pressure to the rods, we must drill passageways in the crankshaft. This crank drilling is no job for an amateur, so we suggest you have it done; a number of shops will do the job for $25. This includes cutting a deep groove around the main bearing journals to provide more oil flow area at this point and to assure full feed to the rods. (The stock Chevy main bearing shells are not thick enough to be grooved.)

The next thing is the oil pump. When connected into a full pressure system, the stock Chevy pump will pull up to 75 lbs./sq. in. pressure and will do very nicely in many applications. For very high oil flow rates induced by large bearing clearances, however, the capacity of the Chevy pump is barely adequate, and we must turn to the high-output GMC 228 unit (see Fig. 4-8). This entire pump unit is interchangeable with that

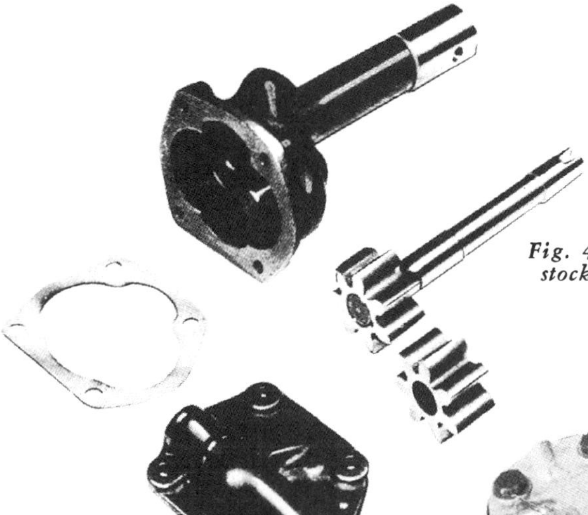

*Fig. 4-7. Parts for the stock Chevy oil pump.*

*Fig. 4-8 GMC 228 oil pump, with shroud and screen removed.*

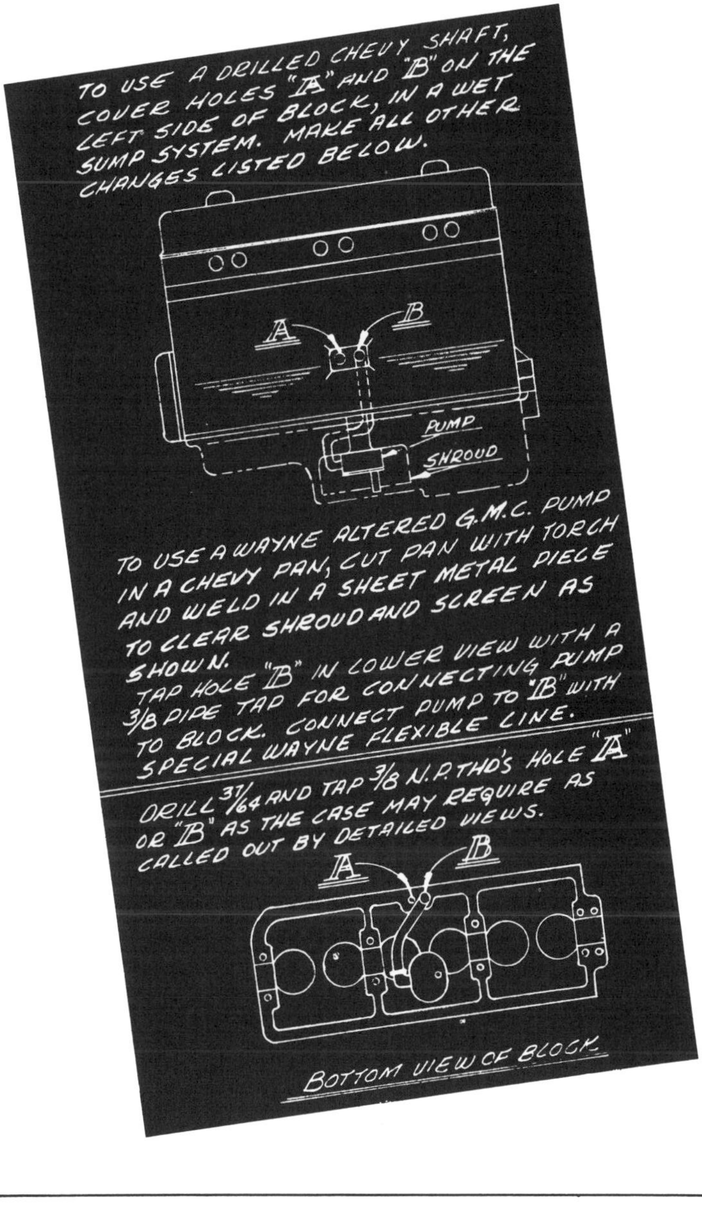


*Fig. 4-9. Wayne drawings showing setup for full pressure wet-sump lubrication system on the Chevy, using no oil filter.*

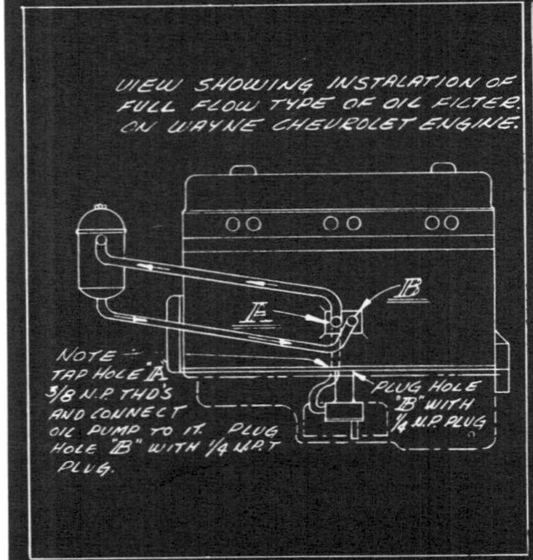


Fig. 4-10. Wayne setup for wet-sump pressure lubrication with full-flow oil filter.

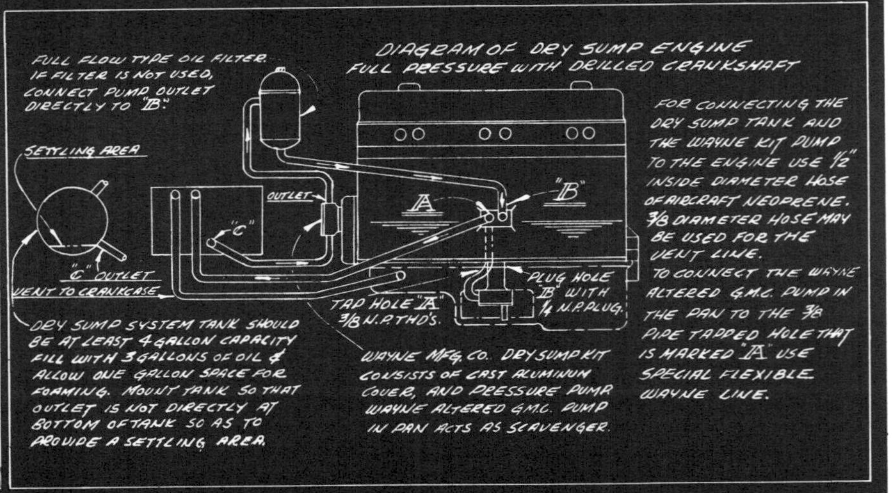


Fig. 4-11. Wayne setup for a dry-sump pressure system on the Chevy, using a full-flow for long-distance track racing at high RPM

on the Chevy and can be installed without changes (except to modify the oil pickup). The GMC pump has a built-in relief valve set to 60 lbs. at the factory, and will take care of the lubrication as is under the most extreme conditions.

Next, rip out the sheet metal tube and piping that feeds the jets in the oil pan. On the side of the block opposite the pump is the oil pressure-regulating and distributing unit; there are two holes in the lower block face at this point (the stock oil pump connecting into the rear hole). Now if you want to use a full-flow oil filter, connect the pump to the *front* hole (see Fig. 4-10) and lead the filter line out of the pressure-regulating unit on the side of the block, and then back into the main oil gallery at this point. If you don't want a filter, block off the regulating unit with a gasket and metal plate and hook the pump into the *rear* hole, so as to feed the oil directly to the main gallery (see Fig. 4-9.) In either case, the block hole you don't use should be plugged with a 1/4-in. pipe plug.

The sheet metal end plugs in the main oil gallery will not hold much pressure; these should be removed and replaced with 3/8-in. pipe plugs (see Wayne instructions in appendix). Wayne suggests using 1/2-in. I.D. Neoprene hose and 3/8-in. pipe fittings for all oil lines, but smaller metal lines could as well be used for road jobs.

The "pressurizing" job is completed by plugging the oil scoops in the rod caps with 1/8-in. plugs. (It is suggested, however, that you invest in special insert-type rods for this job.) Oil feed to the rocker arms is generally left as stock.

Now under sustained high-output conditions as in track racing, oil cooling may become a problem with the stock crankcase capacity of 5 1/2 qts., since it is advisable to keep oil temperature under 200° F. at all

*Fig. 4-12. Installation of a full-flow oil filter on a Wayne-Chevy track engine with wet-sump lubrication system at 60 lbs. pressure. Several engines have run a full season on one set of bearings using the filter.*

*Fig. 4-13. Special Wayne pressure pump for a dry-sump oiling system (the crankcase pump acts only as a scavenge unit); this pump is driven from a special "tang" drive on the end of the camshaft.*

*Fig. 4-14. GMC oil pump fitted to the Chevy block, piped into the rear oil hole to feed oil directly to the main gallery (no filter).*

times. In this case, we have a choice of two remedies: (1) We can enlarge the capacity of the oil pan, or (2) we can go to the "dry sump" system, where the oil is pumped out of the crankcase into a cooling tank (via the stock pump) and then back to the main oil gallery by a special external high-pressure pump.

Figs. 4-15, 16, and 17 show how to build up a 10-qt. "wet" sump with two stock Chevy pans (see also the Wayne instructions in Appendix). We won't explain the dry-sump system, as it is not at all widely used; Fig. 4-13 shows the Wayne pressure pump for dry-sump oiling fitted to the block and driven by the camshaft gear.

## OILS

Stock premium oils will work very nicely in souped engines under all operating conditions. The day when you couldn't hit a half-mile track without castor oil in your sump is past. However, at sustained high RPM in track racing, your bearing loads and heat rejection to the oil are con-

*Fig. 4-15. Parts required to build up a 10-qt. "wet" oil sump, using stock Chevy parts; note horizontal baffle for near section and the vertical baffle.*

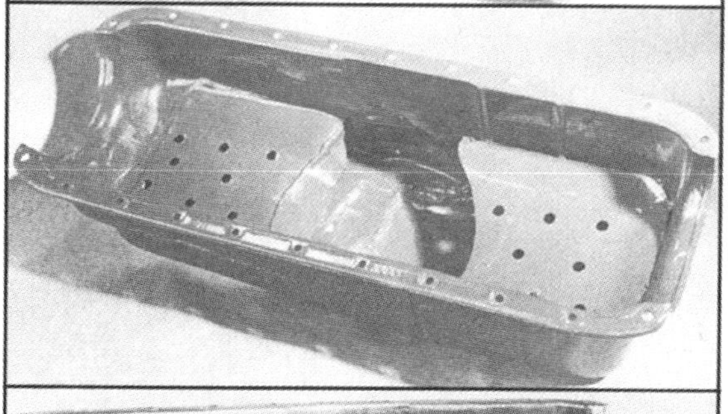

*Fig. 4-16. Ten-quart wet sump assembled from parts shown in Fig. 4-15.*

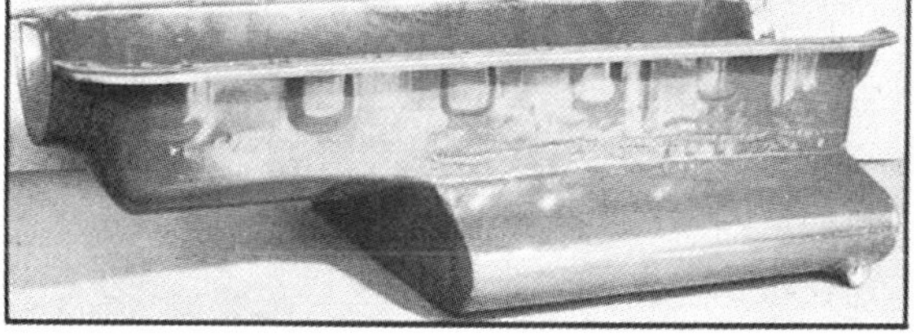

*Fig. 4-17. Special 10-qt. wet sump to clear the steering linkage in late Chevrolet chassis.*

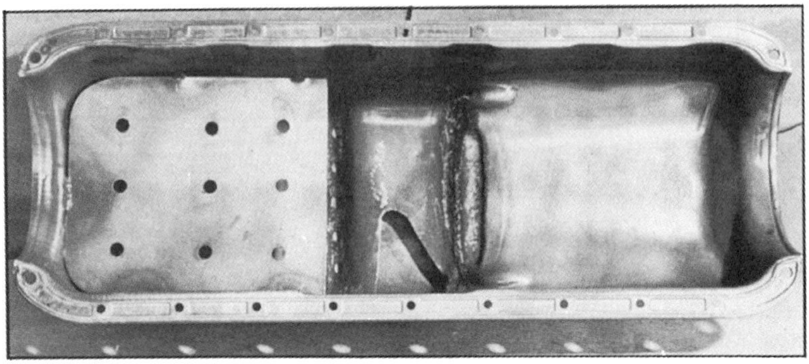

*Fig. 4-18. Stock Chevy oil pan reworked for dry-sump lubrication. Rear section has full baffle (oil troughs removed); slot in pan is to receive the pick-up tube from the pump, which reaches into sump section of the pan; dent across the pan is to clear oil line to the block hole.*

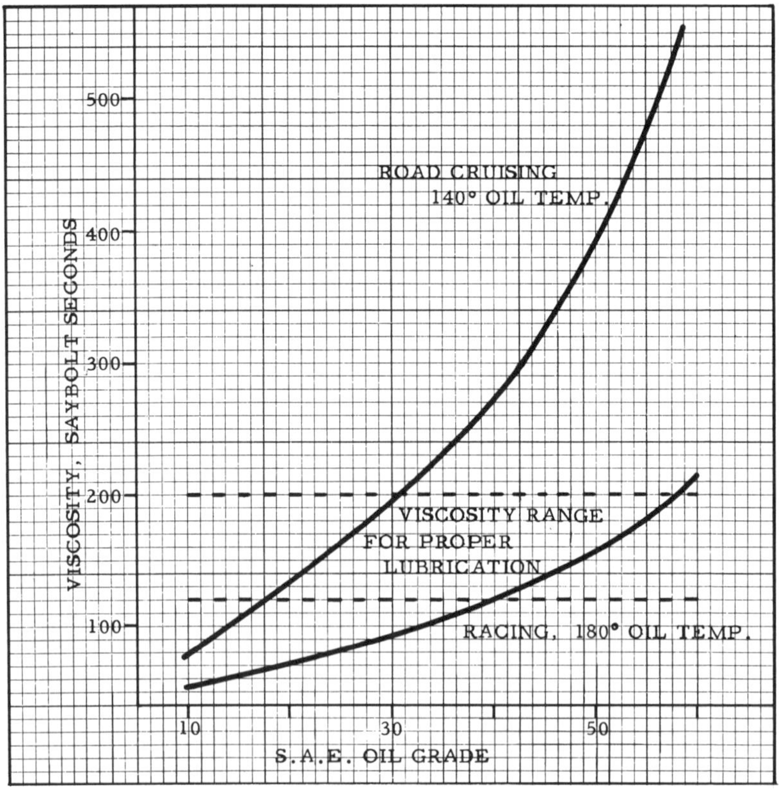


*Fig. 4-19. Graph showing crankcase oil temperature determines our selection of viscosity grade to maintain proper lubrication.*

siderably higher than any stock conditions—which means we must use heavier grades to prevent the oil film from thinning out and "breaking down." The usual practice is to use SAE #40 or 50 for track competition. For road engines, they generally break them in on #20, then go to 30 when oil consumption shows up; for high-speed highway travel in hot weather, use #30 or 40.

We will not recommend any particular oil brands or go into detail on the new "additives." These are matters for experiment and for the recommendations of your engine rebuilding shop.

As for oil filters, these also are a matter for personal opinion and experiment. Wayne definitely recommends a full-flow filter for dirt track racing (the '47 Chrysler 6 type). For road engines, an oil filter won't insure "Rolls-Royce" wearing qualities for the Chevy, but by using the filter, oil never needs to be changed. One very useful piece of filtering equipment, however, is the new Wolfer "CLEAN-OIL" filter for the valve and rocker-arm oil flow (see Fig. 4-20). This is said to eliminate much of the wear in these parts, to reduce noise, and to prevent valve sticking. We definitely recommend the Wolfer filter for any road engine.

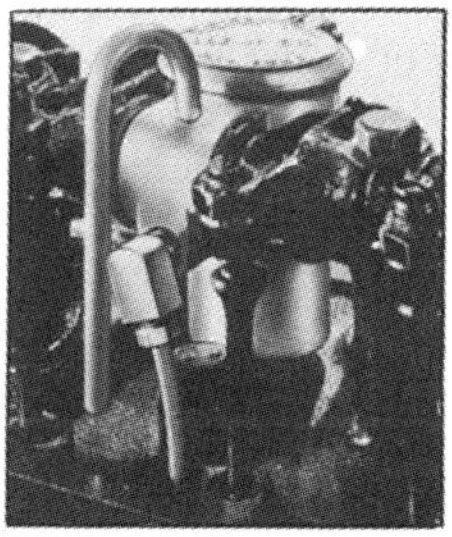

*Fig. 4-20. Right—Wolfer oil filter layout for the valve oil feed.*

*Fig. 4-21. Below—Installation drawing for the Wolfer filter.*

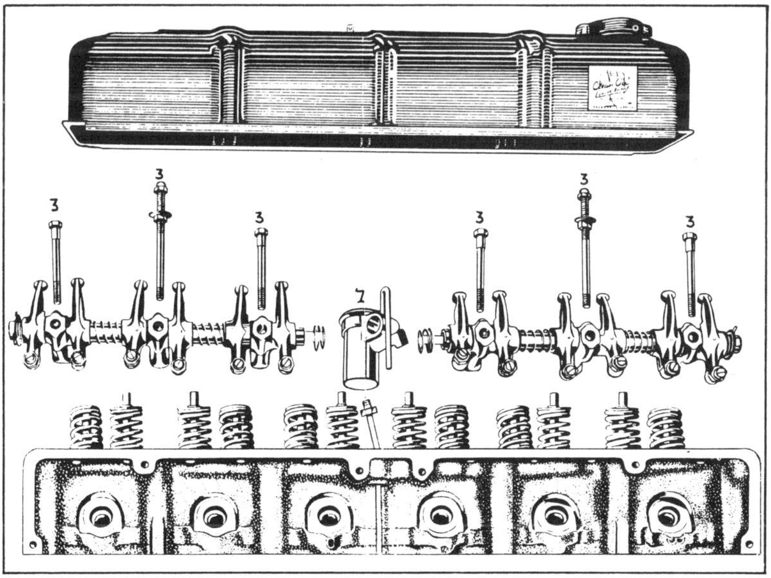

## BEARINGS

Leadfoot Louie doesn't read much technical literature, but when he does, watch out! He was reading something the other day on lubrication theory and he came to the sentence, "Theoretically the bearing surfaces are separated at all times by a solid film of oil and there is no metallic contact."

That hit Louie like a ton of bricks — especially after his recent bitter experience using 0.00025 in. bearing clearances to stop main bearing pounding! Why not use chrome-plated steel linings at a half-thousandth clearance and eliminate all noise and wear? Well, he tried it! He ran the engine just 32.2 seconds — and was able to salvage the head, camshaft, and oil pump (but he never did thaw out that crank train!)

Actually, our bearing theory does say that bearing surfaces are separated by an oil film, with a "wedge" of oil under the point of maximum, load being built up by friction between the oil and rotating shaft. However, there will always be a little metallic contact at times, and this is why you must have a soft lining — to imbed metallic particles worn off (and other impurities in the oil or bearing) and to keep bearing temperatures from rising drastically.

This is all well and good, but there's a catch: In order to get long wear and other desirable bearing features at minimum cost with stock bearings, they must be made of quite soft materials — babbitts, copper-lead, etc. When we double our peak bearing loads, as is the case when we soup the stock engine, these soft linings take quite a pounding, and they are apt to fail from "fatigue" within a few thousand miles under hard driving.

Chevrolet uses a very thick cast lead-base babbitt for the rod linings. Let's face it — this bearing is not good for high loadings. The soft, heavy lining compresses under high load, which ruptures the microscopic bearing structure — and before you know it, you've lost a rod or two. We wouldn't trust the stock Chevy rod bearings much above stock conditions at sustained high RPM (in fact, we have known a good many that wouldn't even take 90 hp on the long grind).

To remedy this situation at a minimum of cost, some handy mechanics fill in part of the big-end by brazing or silver-soldering to within 0.015-0.020 in. over crankpin size, and then rebabbitt in a thin lining. This works very nicely, but it's quite a job and we know of no outfit that does it commercially.

A better solution is the use of stock Chevy rods converted to use precision insert bearings (see Fig. 4-22). Several companies supply these rods at around $30 a set, with or without the oil scoop in the cap (for either stock or pressure lubrication), and fitted with special hard linings adapted to high outputs. We recommend these rods for road engines wherever possible. For full-race engines in the 200-hp range, the stock Chevy rod shanks are not too stout, and the GMC rods should be used, which are already of the insert type.

As for bearing linings, Wayne specifies McQuay-Norris cadmium-base (97%) inserts for rods and mains in their race engines. We'll go along with this and suggest cadmium-nickel or silver linings for outputs of over 180 hp. For most road engines, copper-lead is acceptable.

A word about the all-important matter of bearing clearances. With the jet-scoop oil feed to the rods and the low-pressure feed to the mains on

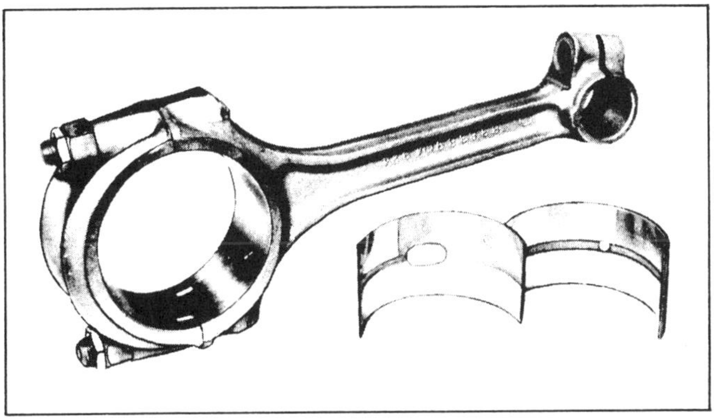

*Fig. 4-22. Insert-type rod for heavy souping. These are stock Chevy rods with big-ends bored out to receive special insert shells, and with oil scoops chopped and plugged.*

our Chevy, the clearances must be right or we are in trouble — more so than on most other engines. (Stock clearances here are about 0.001 in. on the mains and 0.002 on the rods.) Now by increasing these clearances slightly, we increase oil flow rate through the bearing and boost the load it can carry without metallic contact, as well as reducing bearing temperature and friction somewhat. At the same time, our bearing lining is going to take more of a beating, so we have to compromise.

The problem is further complicated by the fact that we *shouldn't use shims on the mains in a souped Chevy.* This means that, to increase bearing clearances, we will have to install new undersize linings and grind the crank journals down to size. This is quite expensive for light souping. If

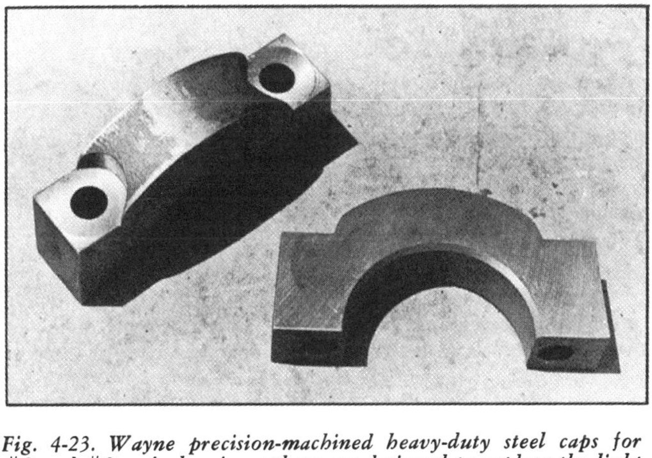

*Fig. 4-23. Wayne precision-machined heavy-duty steel caps for #2 and #3 main bearings; these are designed to replace the light cast-steel stock caps for racing.*

you're building up a hot road engine from scratch, however, we would definitely recommend increased clearances. These should be 0.002-0.0025 in. for the mains, and 0.003-0.0035 for the rods. On a full-race track setup, clearances can go to 0.0035 in. on the mains and 0.0045 for the rods. Camshaft and rocker-arm bearings will be satisfactory in stock form under all conditions.

## THE FLYWHEEL

If it's in decent balance, your flywheel can have *no* effect on HP at any constant speed. Why, then, are we eager to "chop" it? It's all a matter of acceleration. Just as it requires a force to accelerate a mass in a straight line, so it requires a torque to accelerate a mass in rotation. You probably don't realize it, but just the torque required to accelerate the rotating parts of your car's drive line (rods, crankshaft, flywheel, transmission, drive and axle shafts, wheels, etc.) in low gear is equivalent to adding some 40% to the total weight of the car! And the flywheel alone generally accounts for nearly *half* of this. That, in a nutshell, is why we chop them.

Actually, in the Chevrolet, the flywheel is a relatively light 30 lbs. (compared with an average of 40 lbs.). This doesn't leave as much room for improvement as some, but the Chev "wheel" can be turned down on a lathe to around 20 lbs. without any weakening effects, by taking stock off the rounded part of the face. Many shops will do this job for around $10, and it is unquestionably a good investment for any souped Chevy.

(Some claim that what acceleration time you gain on the throttle with a light flywheel, you lose by not getting the "kick" at the shifting point. However, this idea goes against theory, and since the consensus seems to be in favor of the light flywheel for fast acceleration — that's reason enough to chop it.) Incidentally, always have your flywheel rebalanced after chopping.

## COOLING

We are happy to report that there should be "no such animal" as cooling trouble with the Chevy, under any conditions!

## BLOCK REFINEMENTS

In concluding this chapter on the block, we'd like to mention some possible refinements such as honing, chroming, etc. Honing is a means of smoothing up tiny machining marks on the cylinder bore with a lubricated cutter and an abrasive. Opinion is divided on the necessity of it, but we feel that it definitely aids ring seating and should be done if you are reboring. Most shops have a policy on this matter, so consult them.

Chroming is quite another matter. This, of course, refers to hard-chrome plating the cylinder bores and bearing journals. It's a very expensive process. The main idea is to practically eliminate wear in these parts — but a secondary advantage claimed is that friction between the oil film and cylinder or bearing surface is reduced, since the metallic surface would be smoother, providing a bit more HP. Once again, we are unfortunate in having no dynamometer figures with which to prove or disprove the point, but at least the concept is logical and theoretically should make a difference.

However, that difference would be so very slight that we could hardly see going out and laying down $100 or $200 for a full chrome job, expecting to beat all comers at Squeedunk Speedway! Similarly, on the wear angle, why pay $75 to have a crank chromed if you can go out and buy a brand new one for some $35? If you're working with a $250 Offy crankshaft, it is a different matter, but we won't suggest chroming inexpensive stock parts just to save wear. All in all, we suggest you think carefully of all angles before you sink a lot of money in chroming.

On the matter of balancing the crank train, we don't have the problem as with the Ford V8. With a 4-throw crank, your balancing problem is tough, even at the factory. With the 120° spacing of a 6-throw crank, however, the reciprocating inertia forces in one cylinder completely balance those in the next *automatically,* and we need only balance the constant centrifugal forces by crank counterweighting. This means you need worry only about a $40 rebalancing job if you have made extensive modifications with non-stock rods. Even then, a road engine could get by — though we suggest rebalancing a full-race job.

So that about covers the block problem. Once again, we repeat: Be careful — the success of your whole souping job depends on having things set up *right* in the block.

Now let's come up out of the dreary crankcase to more glamorous things. How about the cylinder head? Here's a spot for plenty of souping.

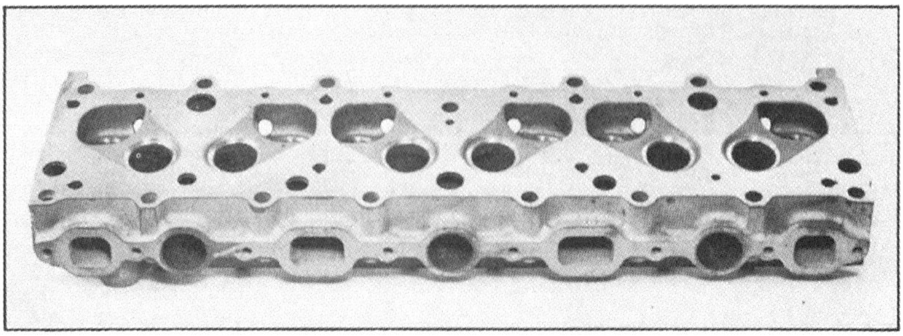

*Special Chevrolet cylinder head for 1937-51 models has 8.1 to 1 compression ratio, has been milled, filled, ported, and relieved.*

*Engine compartment of the "Motor Trend Special," a contestant at Indianapolis, 1951. Special head for 274 cu. in. GMC engine was designed and built by Wayne F. Horning of Los Angeles.*

# CHAPTER 5

# THE CYLINDER HEAD

A LONG time ago, an automotive engineer said, "The vertical overhead-valve cylinder head will save the passenger car engine from an oblivion of low-breathing flat heads."

He wasn't 100% right — but what would he say if he got a look at that new Chrysler head layout? Saints preserve us! Next thing, they'll stick double-overhead cams on our stockers and have done with it.

But anyway you look at it, cylinder head layout is a vital factor in the success of any high-output engine. Leadfoot Louie always thought so too — but Louie usually thinks in a reverse direction. Somebody once told him about valve "reciprocating inertia mass," and for a long time he wouldn't have anything to do with those heavy pushrods and rocker arms associated with overhead valves. He wouldn't have *his* valves floating at 3000 rpm! He's off the L-heads now, though, only because he didn't like the idea of the intake charge having to run *uphill* through the port!

But let's forget Louie's foolishness and look at this head situation in a strictly practical way. On the Chevrolet block, of course, we don't have to choose between side valves or overheads. The overheads are there — and we can be glad they are — for they are one of the fine features of the Chevy engine from a souping standpoint.

But we want to emphasize right here that this subject of gas flow through the ports and valves, which one doesn't usually associate with cylinder head problems, is actually a more important item in our souping than the close-to-home head items like compression ratio, combustion control, cooling, etc. We can often do more for peak HP by improving intake gas flow from the manifold outlet into the cylinder than we can by fiddling with compression ratio and such. Yes, gas flow is the thing — in the cylinder head as well as the manifold.

## GAS FLOW PRINCIPLES

Strictly speaking, our fuel-air mixture coming into the cylinder is not a true gas, but is actually a mixture of a gas (air) and atomized liquid fuel particles in suspension. Less than 10% of the fuel vaporizes to a gas outside the cylinder. However, for all practical purposes, we can treat this intake mixture as a homogeneous gas, and let the laws of fluid dynamics guide us from there.

Basically, this science teaches us that there is a drop in pressure when a liquid or a gas flows through a closed channel — much like the drop in voltage when electric current flows through a wire. This pressure loss is caused by: (1) Friction between the fluid and channel walls and between adjacent layers of the fluid, and (2) turbulence or non-streamline flow in the fluid, which appears as a slight rise in temperature and expansion of that fluid (that is, the pressure energy is converted to useless heat by the work it does in the turbulence).

Of course, we realize what factors contribute to these friction and turbulence losses — rough channel walls, sharp turns, obstructions in the

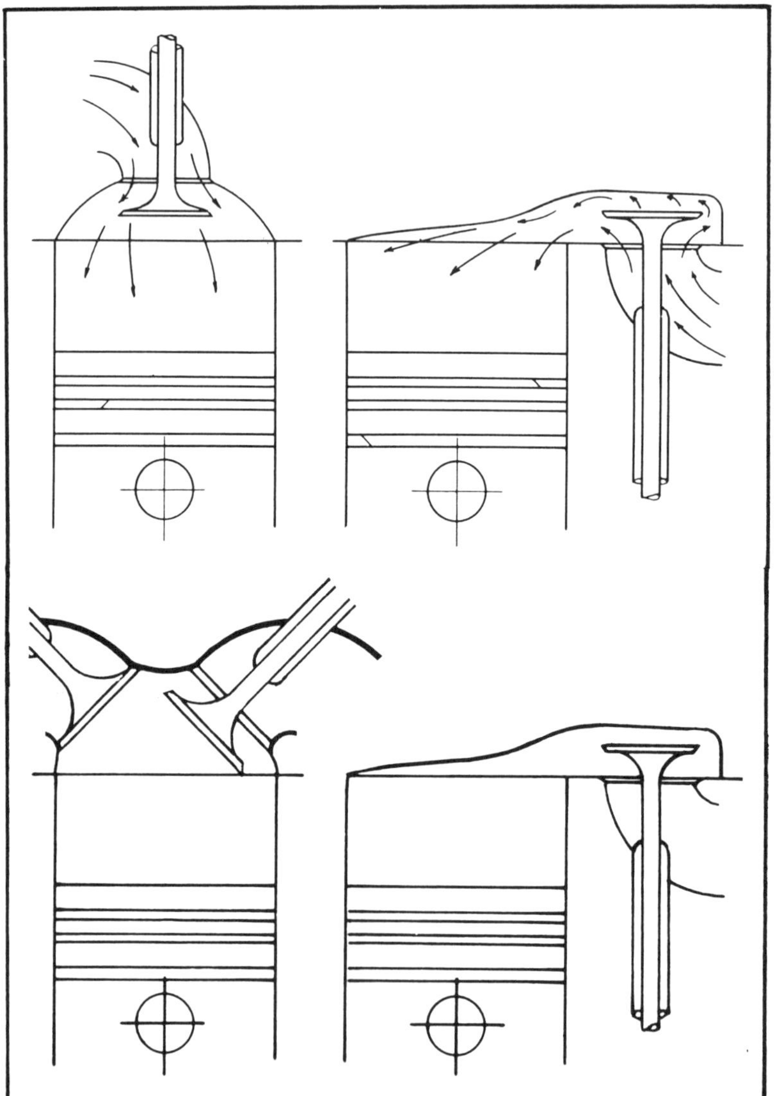

*Fig. 5-1. Comparison of intake gas flow in L-head and overhead-valve combustion chambers.*

flow path, etc. But here is a more obscure, but vital, fact that we should get firmly in mind: Other factors equal, the pressure drop in a fluid flow decreases as the *square* of the flow *velocity.* In other words, suppose we are feeding one cylinder through a port 1 in. in diameter, and getting a pressure drop of 1 lb./sq. in. across the port. Now if we just double the flow path area of that port (increase its diameter to 1.41 in.), we cut the velocity in half — and reduce the pressure loss to ¼ lb./sq. in.

It isn't possible to assess the exact effect of such things as channel roughness, turns, obstructions, etc., on the slide rule, but the idea is evident.

## HEAD BREATHING

When we consider gas flow from the manifold into the cylinder itself, we must consider the flow path from the valve into the cylinder as *part of the over-all channel.* And here's where overhead valves score against the L-heads — that is, they eliminate that rugged, sharp, turbulence-producing 180° turn from the valve port into the cylinder. Fig. 5-1 shows such a gas flow comparison.

The effect of this one little fact on volumetric efficiency is quite remarkable. Some lab tests were run by Cadillac to determine the effect of the 180° turn of an L-head cylinder on its V.E.; the gas flow through the valve at a given pressure was an average of 29% higher with the head off than with it in place, at all flow rates! The effect with overhead valves would be similar to that with the flat head removed. That pretty much proves the point.

Another big advantage of overhead valves is that we can incline them at an angle to the cylinder axis and squeeze 30-40% larger valve diameters into the head (see Fig. 5-1). This means still better breathing. This is a rather embarrassing subject in regard to the Chevy because no one has yet come up with any truly *inclined* overhead equipment for it. This is a pity, too, for it might mean another 20 hp and nearly 6000 rpm under full-race conditions. Something similar to the Ardun overheads for the Ford V8 would be entirely possible on the Chevy block.

So there is a briefing on gas flow as regards the head problem. Remember it, and you won't be filling your head and forgetting to port it.

## COMPRESSION RATIO

We defined compression ratio in Chapter 3 as the *ratio* of the total cylinder volume *at bottom stroke* to the total volume at *top stroke.* In terms of volume, the formula for determining C.R. is:

$$CR = \frac{V_1 + V_2}{V_1}$$

where $V_1$ is the cylinder volume at top stroke (combustion chamber volume) and $V_2$ is the *piston displacement* of the cylinder.

Quite often it is convenient to calculate the compression ratio you are getting with various head changes, and you can use the above formula very easily. The method is as follows: Using the flywheel timing marks, set No. 1 cylinder to exact top dead center and pour thin oil from a measuring beaker into the spark plug hole until it fills the combustion chamber and reaches the bottom of the hole (this means tipping the engine to one side). You can use this measured volume along with the displacement of one cylinder to calculate the C.R. (Remember, if you are using a c.c. beaker, 1 cu. in. = 16.4 c.c.'s.)

Let's work out an example: Suppose you have a standard Chevy block (36.1 cu. in. per cylinder) and you measure the combustion chamber volume as 77 c.c. First we convert this to cu. in., or 77/16.4 = 4.7 cu. in., and the compression ratio is:

$$CR = \frac{4.7 + 36.1}{4.7} = 8.7{:}1$$

Obviously, when we increase the piston displacement by overboring, we automatically raise the C.R. with a given combustion chamber volume. This can have a pronounced effect on such items as knock and spark setting as well as HP. Various head modifications by commercial houses are usually rated as to C.R. on a basis of stock displacement. Below is a table giving true C.R. for general head ratings in terms of the popular bores and strokes:

TABLE I

COMPRESSION RATIO CONVERSION CHART

| BORE & STROKE | RATED RATIO | | | | | |
|---|---|---|---|---|---|---|
| STANDARD CRANK (3¾ in.) | 6.3 | 6.5 | 7.5 | 8.5 | 10.0 | 12.0 |
| Bore 3-1/2 (216 cu. in.) | 6.3 | 6.5 | 7.5 | 8.5 | 10.0 | 12.0 |
| 3-9/16 (225 cu. in.) | 6.5 | 6.7 | 7.8 | 8.8 | 10.3 | 12.4 |
| 3-5/8 (232 cu. in.) | 6.7 | 6.9 | 8.0 | 9.0 | 10.6 | 12.8 |
| POWERGLIDE CRANK (3-15/16) | | | | | | |
| Bore 3-1/2 (227 cu. in.) | 6.6 | 6.8 | 7.8 | 8.9 | 10.4 | 12.5 |
| 3-9/16 (235 cu. in) | 6.8 | 7.0 | 8.1 | 9.1 | 10.8 | 12.9 |
| 3-5/8 (244 cu. in.) | 7.0 | 7.2 | 8.3 | 9.4 | 11.1 | 13.4 |
| 3-11/16 (252 cu. in.) | 7.2 | 7.4 | 8.6 | 9.7 | 11.5 | 13.8 |

Just how does C.R effect HP output? The HP should increase very rapidly as we raise the compression, since we increase the cylinder pressure before the spark fires, which will raise the average effective pressure on the expansion stroke. In the heat and fury of an engine cylinder, things don't quite live up to theory, but increased compression certainly pays off. Fig. 5-2. shows approximately the peak HP increase we get on the Chevy by raising compression ratio from the standard 6½:1, using a cast-iron head (and assuming we burn a fuel that will permit no appreciable knock). We simply add the HP increments shown to our power at stock compression to estimate our true output; for example, if we rework a stock 90-hp Chevy head to 8½:1 compression, Fig. 5-2 shows the approximate power increase to be 13 hp, giving a true output of 90 + 13 = 103 hp.

Notice that the curve levels off at very high compressions, showing that extreme ratios over 15:1 don't do a lot for HP. This is largely because of chemical changes in the combustible mixture at very high pressures and

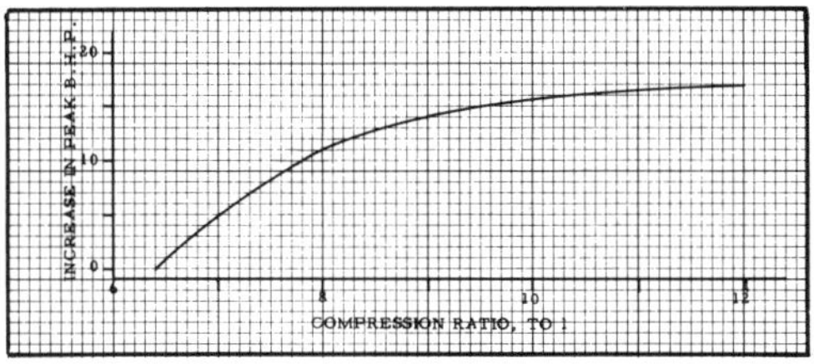


*Fig. 5-2. Graph showing the approximate peak HP increase resulting from boosting compression ratio above the stock 6½:1 (cast iron head).*

temps that absorb more heat in the mixture and hold peak cylinder pressures down.

Another effect of high compression is a reduction in fuel consumption. By expanding the burned gas from a higher pressure, we squeeze more work out of every drop of fuel, and fuel consumption in lbs. per HP per hour drops. Road gas mileage goes up in proportion. In fact, careful tests on a stock '47 Chevrolet running 9:1 compression with alcohol-water Vitameter injection showed an average increase of about 3 miles per gallon over stock or some 18%! Dynamometer test curves on that engine are shown in Fig. 5-3, to further emphasize the great effect of compression

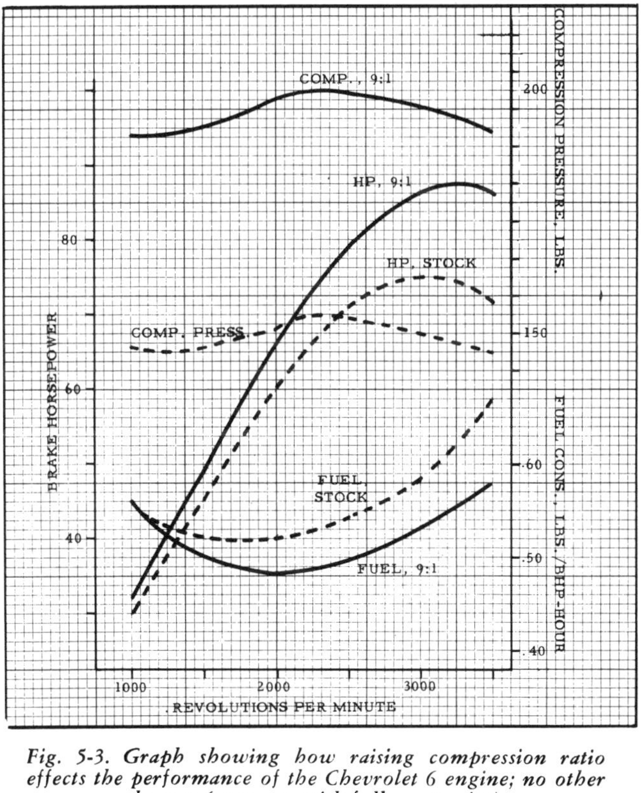


*Fig. 5-3. Graph showing how raising compression ratio effects the performance of the Chevrolet 6 engine; no other changes (tests run with full accessories).*

ratio on performance (the only changes on the engine other than the special pistons were rod bearings and exhaust valves).

Notice also that this souping step is quite effective in boosting HP at low RPM, something like the effect of raising piston displacement. This low-RPM power is vital if we are planning an engine for a lot of city driving—don't forget it.

Lest we get the idea that high compression is always advantageous, let's look at a couple of the bugs: (1) Knock or detonation sets in, and (2) bearings and rings wear faster.

Knock is the most serious drawback here. This phenomenon is merely an instantaneous explosion of the unburned portion of the mixture ahead of the normal flame front, caused by excessive heat and pressure. This sets up violent shock waves in the cylinder, and if severe enough, will quickly wreck an engine by literally pounding it to pieces. When we increase C.R., it's obvious that we increase the tendency of the fuel to knock because of the higher cylinder pressures and temperatures. On the Chevy, using premium (Ethyl) pump gas, severe knock under high load usually sets in over about 7½:1 compression.

It is difficult to give definite figures on bearing and ring wear. Obviously the extra pounding of the higher cylinder pressures should theoretically raise hob with engine parts, but it's not quite as bad as it looks. With 7½-9:1 compression, you can figure on perhaps 50% faster wear on rod and main bearings and rings.

Now let's see exactly what we can do with the cylinder-head problem on the Chevrolet more specifically.

## REWORKING STOCK HEADS

With the overhead-valve layout on the Chevy, "porting" becomes part of the overall head job, as well as increasing compression ratio (however, we leave the valve gear question for the next chapter).

Leadfoot Louie solved his compression problem in one prodigious boner! He thought he could boost compression ratio by packing wet fire-brick clay into the combustion chambers, and depend on combustion heat to glaze the job. He sure had things gummed up for a while! But seriously, we can do wonders with the stock Chevy head in the way of increased compression and port area with some smart modification.

Consider first raising compression ratio. We can do this in either or both of two ways: (1) Milling off the bottom of the head and (2) filling

*Milling*—Milling is the simplest job, and many shops will do it for around $30. On the Chevy head, however, this will have to include sinking the intake valve seat the same amount that is milled off. This "sinking" is done with a spot facing tool and is not a complicated job; however, if the shop that does the job has had no experience with it, be sure and specify that the seat be sunk sufficiently so that the valve head, when seated, will come at least 0.010 in. back from the face of the head. Another point to remember: To maintain the stock valve spring tension after you have sunk the valve, you must insert shims under the spring washer next to the head; shim thickness should be the same as the amount you have sunk the valve. Also, it's a good idea (though not essential) to shim up the rocker shaft brackets when milling.

As for the maximum amount to mill off, this will vary. Actually, you can take off over 0.150 in. on any of the heads without seriously weakening the sections, although that is not advisable. However, the 1937-'40 engines used dome pistons, so that about 0.060 in. is the maximum cut without redoming the head — which is a considerably more expensive job. (This 0.060 cut will raise compression ratio to around 6.8:1 on the 1937-'40 engines.) Incidentally, if you do redome, remember that head clearance over the piston should be at least 0.030 in.

Some have tried to get more compression without having to redome by milling the head "cockeyed" — that is, cutting down to around 0.060 in. on the intake-valve side and 0.090 in the larger space under the spark plug. We don't recommend this. You'll have trouble lining the head bolts, the pushrod cover, and the gasket fit is likely to be poor when the engine heats up under high load. So mill it straight!

One other possibility with the older dome-piston engines is to install new flat pistons; these are interchangeable and will permit much larger mill cuts on the head surface — up to at least 0.135 in. On the '41 and later engines, with flat pistons, your cut is limited only by section thicknesses —

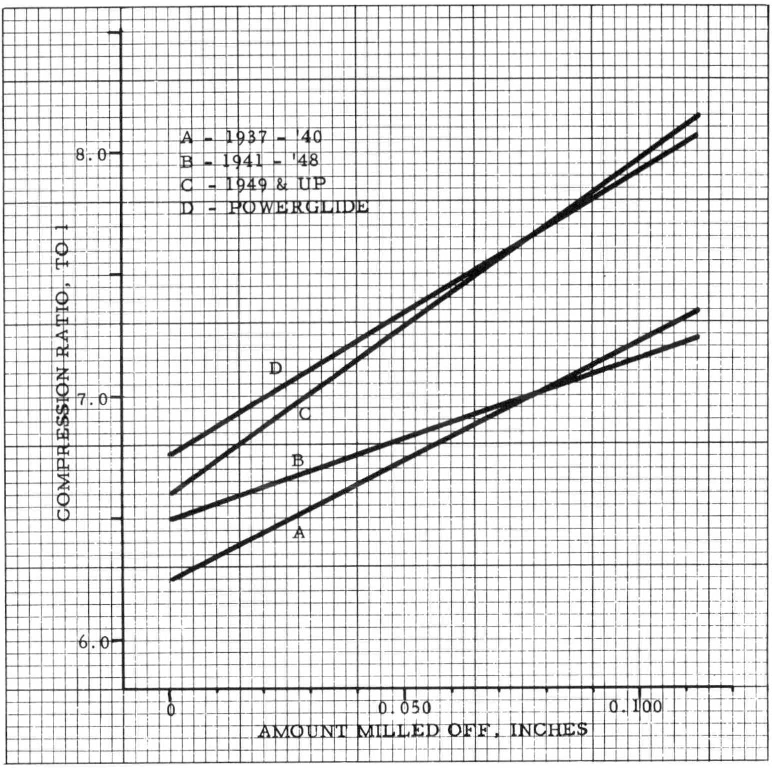


*Fig. 5-4. Approximate compression ratio obtained by milling the various Chevrolet heads (not including filling); research figures courtesy HOT ROD magazine.*

For example, suppose you had the stock Chevy bore and stroke (36.1 cu. in. per cylinder), you were going to mill an early head 0.125 in., and you wanted a final compression of 11:1. The milling would give about 7.2:1 (see Fig. 5-4) and combustion chamber volume would be: $36.1/7.2 - 1 = 5.82$ cu. in. To get 11:1, the volume would have to be reduced to: which, however, are a bit less than on the early heads. We recommend not over 1/8 in. (0.125) be milled from the late heads. Fig. 5-4 shows approximately what compression ratio you get with a given mill cut on the various heads (no other changes).

There are still a couple of other vexing matters to consider when we mill the head — that is, shortening the tappet base-to-rocker length and shortening the pushrod cover panel. (These are all things we need not think about when we shave a flat head.) Fortunately, the Chevy has a very broad tappet adjustment range, so we can take up the amount milled off here without touching the pushrods themselves. And also, shimming up the rocker shaft brackets helps, too.

The push rod cover panel will require some modification, however — and the more you have milled off, the more careful you should be to prevent oil leaks. The panel modification can be accomplished by merely "slotting" or lengthening out the screw holes on the *block side* in an upward direction to take up the amount milled off. This job can be done with a hand grinder, drill press, or hack saw. Be sure not to slot the holes for the head screws, as these must retain the plate in the proper position (also remember that the gasket will have to be slotted.) On the deeper mill jobs, you may also have to cut off the lower edge of the panel to clear engine parts. All this is not a tough job, but be careful, or you may end up with a bad oil leak. On the Powerglide engines, this operation will not be necessary, as the panel does not extend onto the head.

*Filling*—As its name implies, filling consists of filling in sections of the combustion chamber with metal to raise the compression ratio. It is the only method by which we can get much over 8:1 on the Chevy head — and, in fact, we can go up to 12:1 quite nicely this way.

Procedures for doing the job vary. In general, the head is stripped, heated in a furnace or open charcoal or firebrick bin (or merely with a torch ) to a cherry red heat, about 1,500 F., and the combustion chambers are then filled in, using an acetylene torch and cast iron welding rod, while the head is red hot. When the head has cooled, the "fill" is ground to smooth it up and the head is surfaced off to give a level face, since the heating *is bound to warp it.*

There are a number of shops that will do this job, but if you decide to try it yourself, here are a few tips: Heat and cool the head *slowly* to prevent cracking. Decide what compression ratio you want first, what volume will be milled off afterward, and then calculate the volume of fill needed. You can use this formula:

$$V_1 = \frac{V_2}{CR - 1}$$

where $V_1$ is the combustion chamber volume, $V_2$ is the cylinder displacement, and "CR" is the compression ratio.

36.1/11 – 1 = 3.61 cu. in. Therefore you would have to fill in about 5.82 – 3.61 = 2.21 cu. in.

The usual welding rod runs 3/16 in. in diameter, or 0.028 cu. in. per inch of length – so you should run 2.21/0.028 = 79 in. of the rod into each chamber. Be sure to fill each chamber the same amount and in the same shape. The Chevy chamber is shaped somewhat like an "L", with the exhaust valve at the center; the usual practice is to fill in the ends of the "L" right up to the exhaust valve and spark plug (see Fig. 5-5).

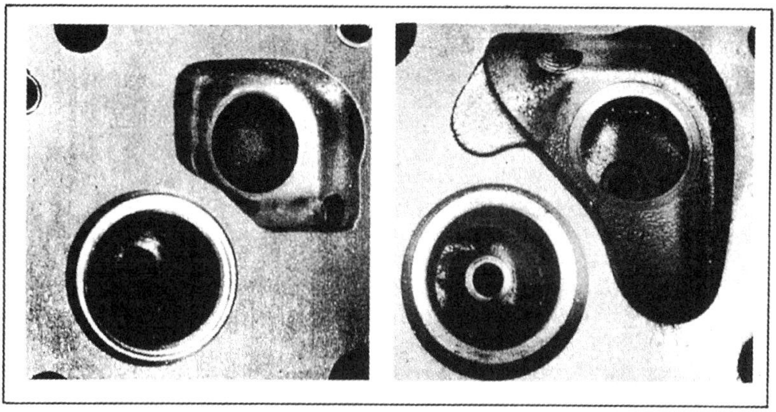

*Fig. 5-5. Photo illustrating "filling" The combustion chamber on the right is the stock form with 6½:1 compression ratio; chamber milled to 10:1 compression.*

After you have filled the head and ground the fill out to smooth it up, check the chamber volumes with oil and the measuring beaker. If some volumes are off (which they are apt to be), grind them out to equal the largest volume of the six cylinders. When the job is finished, surface off the head to allow for warping. By combining these methods of milling and filling as outlined, you can get up to 12:1 compression – which is certainly high enough for most purposes.

We can't help but state that this business of filling *looks* horrible from a technical standpoint. Theoretically the engine should knock like mad on pump gas at 9:1, where aluminum Ford heads are operating beautifully. In practice, though, it is not as bad as it looks, because the maximum length of flame travel is only a couple of inches – but it remains that knock is more critical with a filled head. So if you can get your compression any other way, don't fill!

## PORTING

The Chevy head has always been difficult to "port." That's because the wall sections in the head casting are unusually thin, making it dangerous to do a lot of grinding or reaming. However, as we mentioned, the job is very desirable from a breathing standpoint, and in fact, it's worth 5 or 10 hp at the top end! It can be done. Here's how:

The valve port diameter on the standard Chevy head is 1.25 in.; this

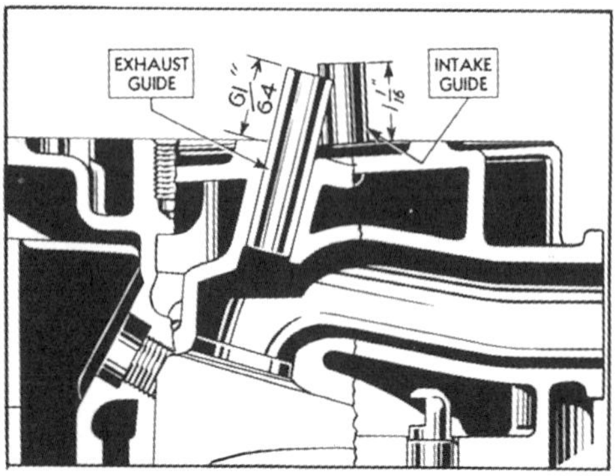


*Fig. 5-6. Stock Chevy cylinder head layout, showing arrangement of porting, valve guides, etc.*

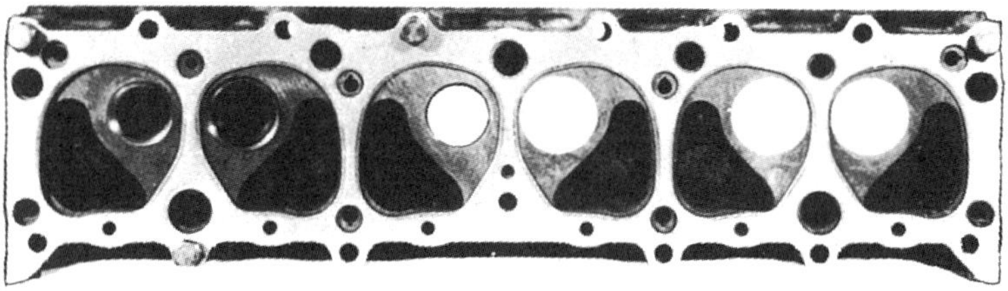

*Fig. 5-7. Photos illustrating valve porting. The far left cylinder is stock; No. 2 intake port has been bored out and refaced to take a 1-7/8" intake valve, and the other cylinders to the right have the oversize valves fitted.*

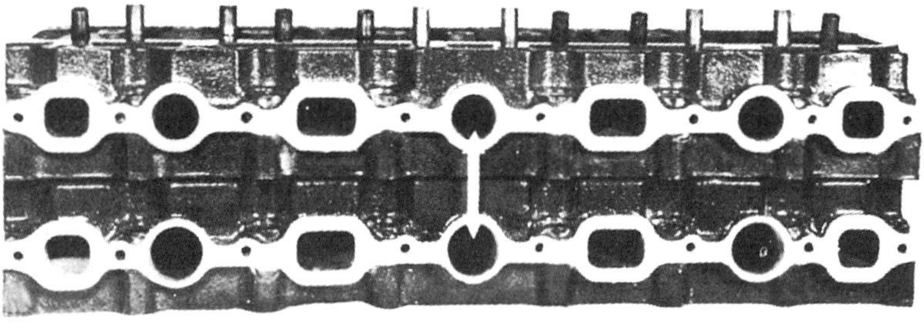

*Fig. 5-8. Photo illustrating head porting; intake port openings in the lower head have been opened out as specified in the text (top head is stock).*

can be decreased to 1-3/8 in. on 1941 and later heads, and to 1-7/16 on the 1937-'40 head. The exhaust ports can be opened out some, but since exhaust gas flow is under considerable positive pressure, detail refinements here have little effect. A shell or sleeve-type reamer setup is needed to do this job, mounted on an arbor with a pilot shaft to extend up into the valve guide to align the ream. This pilot should have at least 0.003-0.004 in.

clearance in the guide, or that is, a diameter of 0.340 in. The reamer should be run right up under the guide.

On the late Powerglide heads, the stock intake port diameter has been enlarged to 1.44 in., and this can be further opened to 1-5/8 by reaming. Exhaust ports are usually left untouched.

The three intake ports on the side of the standard head can be ported out to about 1/16 in. over the stock diameter of 1-3/8 in. (or to about 1-7/16 in.). These ports go straight into the casting, so you can just stick the head in a drill press and start reaming; where the ream meets the "Siamese" branch, it must be smoothed up with a hand die grinder. On the Powerglide heads, you can port out the side intake ports to 1-1/2 in. with safety. Exhaust ports on both heads are curved, and are usually left alone in the porting job. Before going on, let us emphasize once again that sections are thin in the Chevy head, so don't go above these recommended diameters.

Remember when porting that you should also port the manifold the same amount that the side holes on the head have been opened. The gasket should also be ported. These last points are simple, but they are *just as important* in getting that last HP as any of the work on the head.

Obviously all this head work is a big, and rather complicated job. It's really not a job for an amateur—especially in view of the thin sections with which we are working. If you didn't want to tackle the job yourself, or to pay the little machine shop around the corner $100 to do it for you, we wouldn't blame you. Fortunately for the amateur Chevy souper a number of companies (mostly in California) rework the Chevy head on a production basis and sell them through the mail.

These are stock Chevy heads, milled, filled, with intake valves sunk, sometimes with larger valves fitted, new guides, drilled and tapped for 14-mm. plugs, ported clear through, and available in compressions from 7:3 to 12:1 (see Fig. 5-9.) Prices (with your old head) range from $50 to $90. These heads are very fine for road engines, though unless unless

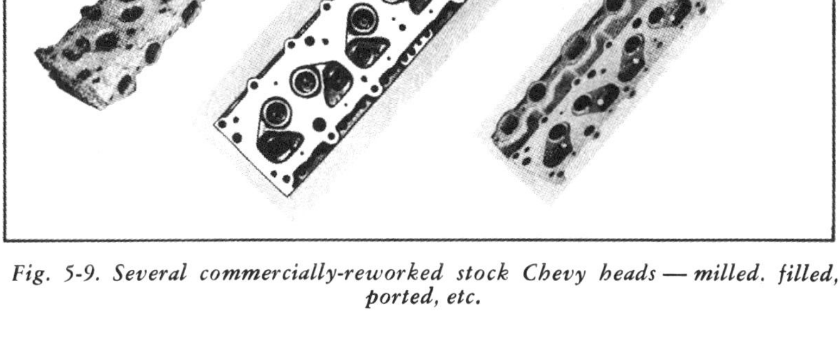

*Fig. 5-9. Several commercially-reworked stock Chevy heads — milled. filled, ported, etc.*

you are nearly broke, they are a killing compromise for racing. A popular ratio for the road is 7.7:1, and this seems to work nicely without too much knock. We strongly recommend these reworked heads as a good investment for a Chevy road engine—but don't expect miracles from them on the track—even on alcohol at 12:1.

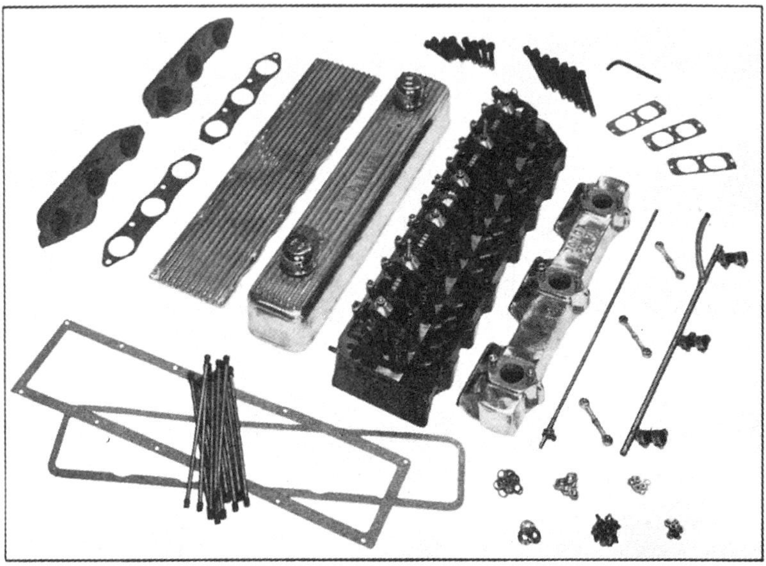

*Fig. 5-10. Complete Wayne-Chevrolet head assembly with exhaust manifolds, pushrods, etc.*

*Fig. 5-11. Wayne head set assembly.*

## THE WAYNE HEAD

The "big stuff" in the Chevy field as of now, is the "Wayne" head, built by the Wayne Mfg. Co. of La Crescenta, Calif. It is the only special racing head for the Chevy block being produced for public sale. It was originally designed by Wayne Horning and Hary Warner, but the business was recently bought out by Warner, who is now supplying complete Chevrolet speed parts as well as complete racing engines built up with the Wayne head and GMC parts. (Horning has opened another company specializing only in the GMC block.)

The Wayne head is a cast iron rocker-overhead outfit with two vertical valves per cylinder and six individual intake and exhaust ports on opposite sides of the head. Figs. 5-10, 11, 12, 13, and 14 show the layout. Compression is varied by using different piston crowns with the special Venolia pistons (the available range being 10 to 14:1). The whole layout is beautiful from a HP standpoint, with a smooth, round, fully-machined combustion chamber, and large valves and ports for maximum breathing.

Careful attention has also been given to providing huge water spaces, especially around the valves, to assure ample cooling under racing conditions. Buick rocker arms of 1½:1 lift ratio are utilized in conjunction with '47 Cadillac valves, cut to 1.75 in. intake and 1.62 exhaust. Stock Chevy

*Fig. 5-12. Intake side of the Wayne head.*

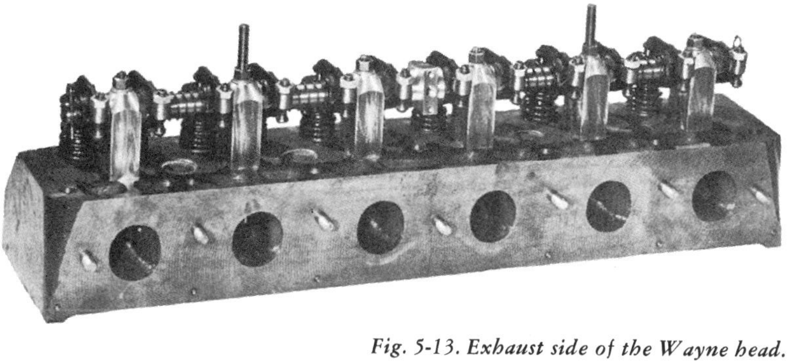

*Fig. 5-13. Exhaust side of the Wayne head.*

*Fig. 5-14. View looking up into the smooth, round combustion chambers of the Wayne head; note large valves.*

valve springs and retainers are used, but a special inner spring is added to boost the total tension to 250 lbs. with valve open. Otherwise the Wayne head is special throughout. (Instructions for installation are given in the Apendix.)

The Wayne head is definitely terrific for the Chevy, both dollar and performance-wise. Over 260 hp at 5000 rpm has been obtained with the full setup—without nitro! We feel they have compromised with economics by not inclining the valves, but the fact remains that the thing does *perform.* It's a must for the man who wants to race — and win — with his Chevy.

That's the story on the chevy cylinder head. It's a cinch that there is more to the problem than whacking off fifty-thousandths, as we can on the Ford flats. But it's also certain that we can't get a more practical extra 20 hp, and pave the way for hot cams and carbs, *easier* on any other stock engine than we can by scientific modification of the Chevy head. If this reworking isn't enough for you, you can always turn to the Wayne head and jump into the 200-hp bracket. Yes, the head problem is well under control in the Chevrolet souping field.

Now let's take the next logical step on the souping trail, the induction system — where horses come in 10's instead of 5's at the top end — and where low-speed virtues seem to disappear!

# CHAPTER 6

## THE INDUCTION SYSTEM

HERE is one for the books: Can you imagine trying to get any performance from an engine if you had to open the intake valve by *cylinder vacuum*? But that's the way they did it 50 or 60 years ago!

Then someone came along and set the automotive world on its ear by opening and closing the valve with a cam and spring. Big stuff! It wasn't long afterwards that they started putting the valve overhead for better breathing. Then someone else inclined the intake valves and put two of them in each cylinder.

Soon the supercharger burst upon the scene. At last — pressure in place of vacuum! In the early '20s, Europeans came up with still another bold stroke that is helping us still: They found they could pack more charge into the cylinder by utilizing the cooling effect of alcohol. Since the last war, big things have been done with fuel injection and weird, radical cam timings.

And what's the latest? Super fuels, of course. They are washing the oxygen through the carb jets these days. We wonder what will happen to the lowly intake valve if this keeps up — probably be tossed out altogether and explosives injected right into the cylinder *without* air! And where does all this leave Leadfoot Louie? Oh, he's still worrying about whether his mixture should run uphill or downhill in the ports!

This matter of the induction system is definitely the most profitable point of attack in souping any stock engine — the Chevy included. The stock designers have given us very little gas flow refinement to start with, so there's a lot we can do. And the reward will be a flock of cheap horses at the top end of the RPM scale — grief at the low end.

### FUELS

In the first place, we will not go into the subject of "super fuels" here. These are merely explosive compounds that carry their own oxygen — which, of course, alters the entire problem from a mixture-induction standpoint. Besides, at this time, the future of super fuels in competition is in doubt, and since they are not practical for road use, we feel the problem is not of definite significance.

So if we consider only the conventional fuels which require air to burn them, we find that the field of possible fuels is narrowed considerably. For a road engine, of course, we must burn gasoline that is available at the pump, so we have no choice here at all — except to remember to keep compression ratio below about 8:1 to prevent excessive knock under high load.

For competition, however, we are not limited to pump gas and can select the fuel that will give the best overall performance within cost limits. Forgetting costs for the time being, we find three main factors governing our fuel selection: (1) Octane rating, (2) volatility, or tendency to evaporate, and (3) latent heat of vaporization.

Consider octane requirements. We won't go into the technical explanation of octane numbers here, but suffice it to say that this value indicates a fuel's relative resistance to knock or pre-ignition. For example, premium pump gas rates about 80-octane with lean mixture, and 90 rich; benzol runs 87-120; alcohol 90-200 (all depending on mixture richness). Now our octane requirements under high-output conditions in road or track racing with 9 to 14:1 compression — that is, the fuel octane we will need to prevent wrecking the engine from continuous knock or pre-ignition — will range generally from 90 to 120 (see Fig. 6-1).

Obviously pump gas is not satisfactory here. If cost is a big factor, 50% benzol mixed with pump gas will give the required knock resistance in most cases; otherwise 91-octane aviation gas (at around 35-40c a gallon) or the more expensive fuels will be fine. A word of caution: Never use

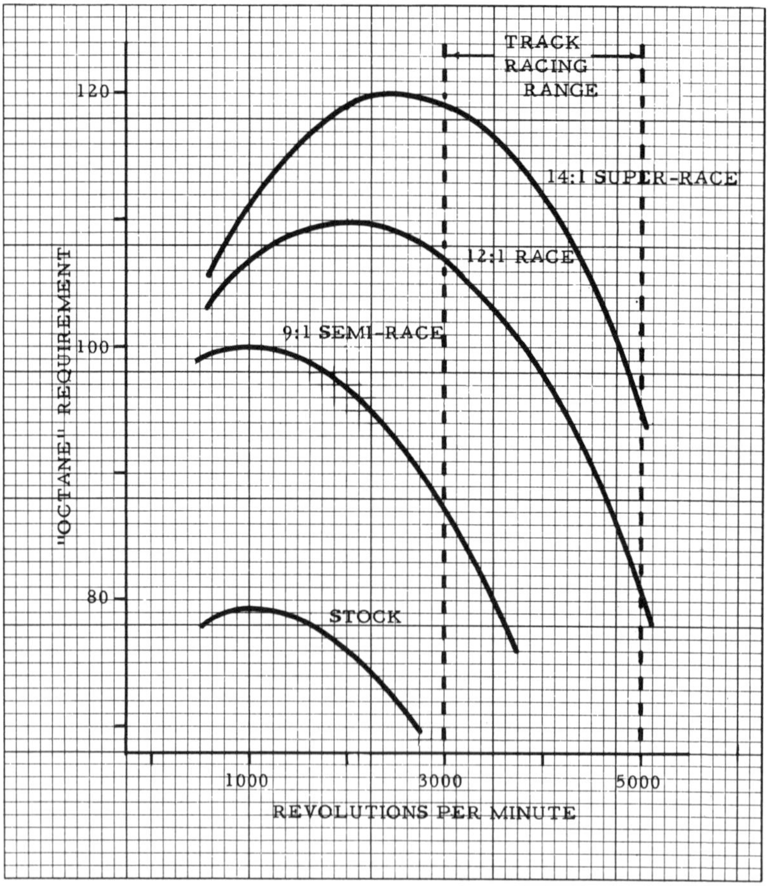


*Fig. 6-1. Approximate fuel octane requirements (full throttle) with various degrees of souping on the Chevy block; cast iron head.*

super-octane aviation fuels, such as 130-grade combat gas, in a souped stock engine; these burn very slowly and will always overheat.

Consider volatility. This isn't too important in track racing where there is no low-RPM acceleration, but in road racing we must consider it to assure good mixture distribution coming out of corners at low speed. Here is a list of the usual racing fuels in the order of their volatility:

1. Ether
2. Acetone
3. Gasoline
4. Methanol (wood alcohol)
5. Benzol
6. Ethanol (grain alcohol)

Because of this matter of volatility, you can't very well use pure alcohol for road racing, but rodders often mix in 25-50% gasoline, or sometimes 5% or less of ether or acetone. Incidentally, because of its low volatility and high weight, alcohol tends to be "centrifuged out" in a supercharger, which floods the casing and starves the engine; don't try using more than 85% alcohol in blending fuel for a blown engine.

From the standpoint of pure HP, the most important characteristic of a fuel is its "latent heat of vaporization." As we know, any liquid absorbs heat when it evaporates and expands to a gas. Let some alcohol evaporate from your hand and notice how cold the spot feels. Similarly, when the liquid fuel vaporizes in the manifold, it absorbs heat from the air in the mixture—which contracts it so that we draw a greater *weight* of mixture into the cylinder on the intake stroke. And our HP shoots up in proportion!

Fuels vary widely in the amount of heat they absorb when evaporating, and this amount of heat, measured in B.T.U.'s per lb., is called the latent heat of vaporization. Here is a list of relative values for the various fuels:

1. Methanol, 500
2. Ethanol, 390
3. Acetone, 235
4. Benzol, 162
5. Ether, 160
6. Gasoline, 135

This is why alcohol will show a peak HP increase of 10% or more over gasoline without any other changes. As a result, pure methanol is absolutely the best bet for maximum possible HP. The disadvantage, of course, is the very high fuel consumption with alcohols. When we run a long-distance race, the pit stops for fuel might cost more in average speed than the 10% less HP with gasoline. But for short sprints, you can't beat methanol without getting into the super fuels.

Let's sum it up thus: If you are not going all-out with your racing, and if fuel cost is a big factor with you, a mixture of pump gas and benzol or 91-octane aviation gas will do. If you are serious about sprint competition, fit up for pure methanol (NOT ethanol). For long-distance racing, gasoline alone will give the best fuel tank mileage, but you should weigh other factors such as cost and HP before making a final decision. (We might mention here that the favorite mixture for the 500-mile Indianapolis race

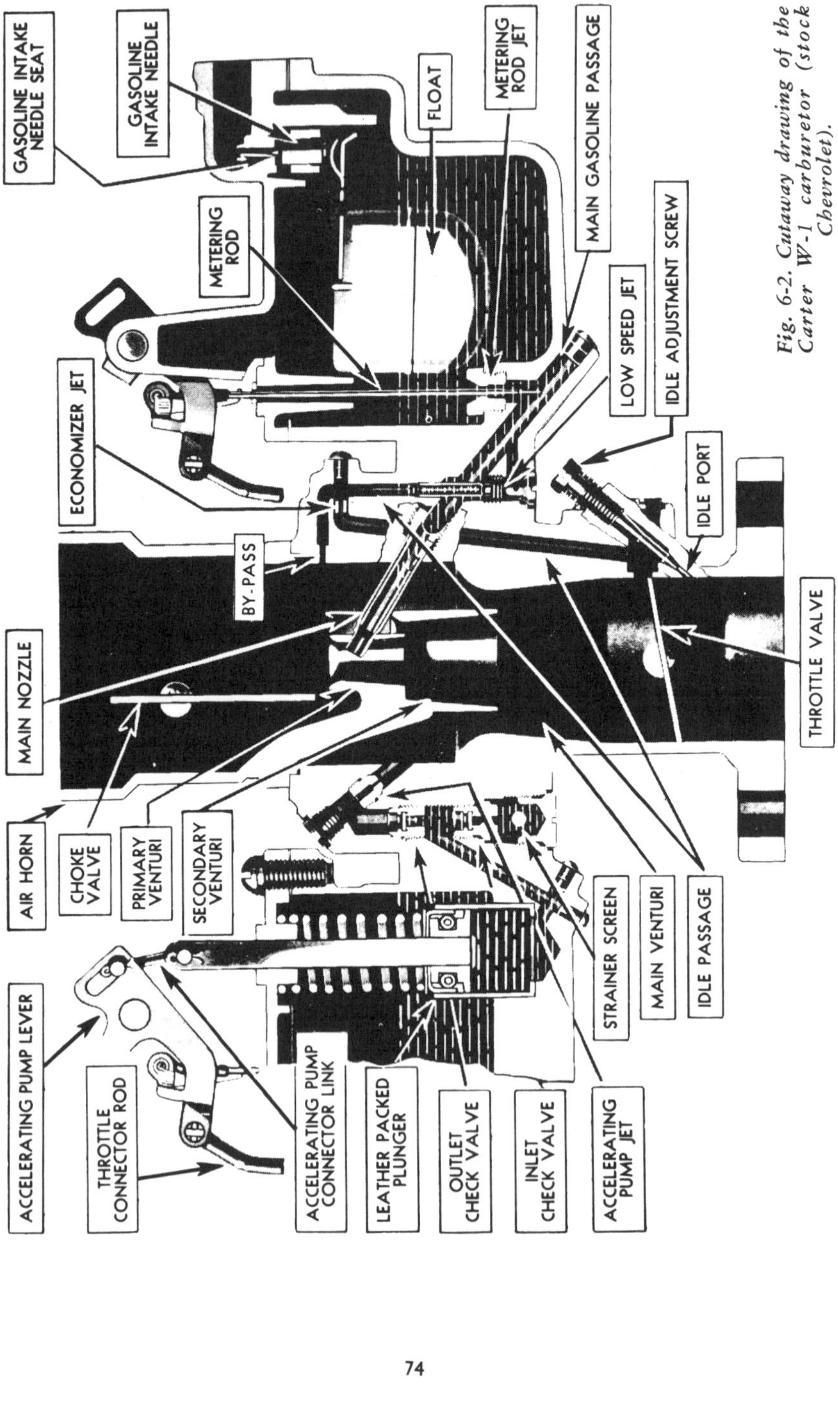


*Fig. 6-2. Cutaway drawing of the Carter W-1 carburetor (stock Chevrolet).*

is 40% methanol, 40% gasoline, and 20% benzol.) For road racing, gasoline is good for the needed volatility, but if you mix over about 60% alcohol, better figure on a small amount of ether.

## CARBURETORS

The carburetor is probably the least of our worries in this souping business. The stock units will generally work very nicely under all conditions without changes. However, some carb models are considerably better than others from a standpoint of flow losses, which, of course, can effect the overall performance of the engine a lot.

The most important factor here is the choke tube diameter. This should be large for the minimum flow resistance at high flow rates (high RPM); for best mixture distribution and acceleration at low speed, the venturi should be relatively *smaller* to give higher gas velocities. We face a compromise here, so we must base our selection on the particular performance requirements.

Internal pressure losses resulting from the design of the carb (obstructions, venturi shape, etc.) are less important, but they do have their effect, especially at high speed. Of the carb brands that figure in the Chevrolet souping picture, this is about the way they line up, those with the lowest pressure loss at the top:

1. Zenith
2. Stromberg
3. Carter
4. Rochester

From this point the problem is only one of getting the right fuel metering characteristics and hooking up. Actually, the stock metering setup will work very nicely with multiple carburetion. These carbs are designed to supply the mixture ratio according to the air flow through the venturi; when this flow is halved by using two carbs, they simply meter less fuel to compensate.

However, we can sometimes get a little more HP by altering jet or metering-rod size a bit to the rich side (which also helps the cooling)—or we can go the other way to the lean side and increase gas mileage, at the cost of a few HP and hotter running. It is a matter of experiment, though, as individual engines and carbs vary (we knew a fellow who went "rich" a little and cut his top speed by 15 mph!)

There are a number of different carb models that can be used on the various dual or triple Chevrolet manifolds now on the market. The most popular models:

*Carter W-1*—This carb was standard equipment on all Chevys up to 1949, and has a nominal 1.25 in. barrel size (single down-draft type). Fig. 6-2 shows the layout. The main metering system is by a tapered metering rod in a jet, which makes this model rather difficult to work with and to adjust under various competition conditions. Metering rod-jet combinations are available one step lean and one step rich, and the lean step is generally specified for setting up multiple carburetion (though only as a step toward decreasing fuel consumption).

*Carter WA-1*—This is very similar to the W-1, except barrel size is 1.50 in. It has the disadvantage of the metering-rod setup, but is considerably

## PARTS ILLUSTRATED

1. Choke Valve
2. Choke Relief Valve
3. Vacuum Power Piston
4. Main Discharge Jet
5. High Speed Bleeder
6. Idle Tube
7. Float
8. Float Needle Valve
9. Float Needle Valve Seat
10. Float Fulcrum Pin Clip
11. Float Fulcrum Pin
12. Power By-Pass Jet
13. Main Jet Plug
14. Main Metering Jet
15. Throttle Valve
16. Idle Discharge Holes
17. Idle Needle Valve
18. Vacuum Spark Conn.
18A. Vacuum Spark Hole
19. Primary Venturi
20. Auxiliary Venturi
21. Idle Air Bleed
22. Float Chamber Vent
23. Piston Conn. Rod
24. Cotter Pin
25. Pump Piston Rod
26. Pump Duration Spring
27. Pump Discharge Nozzle
28. Pump Piston
29. Pump Piston Exp. Spring
30. Pump Relief Valve
31. Pump Inlet Strainer
32. Pump Link
33. Pump Lever
34. Pump Inlet Plug
35. Pump Inlet Check Valve
36. Pump By-Pass Jet
37. Throttle Stop Screw

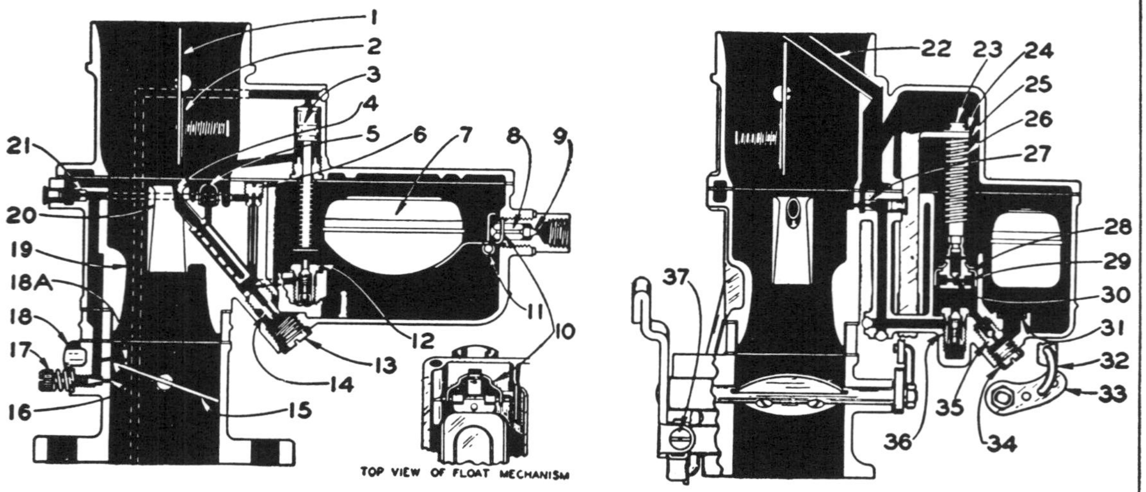


*Fig. 6-3. Sectional drawing of the Stromberg BXOV-2 carb with parts list; note lack of obstructions in the air flow path and accessibility of the main jet.*

better than the above for high-RPM work.

*Stromberg BXOV-2* — This is a replacement unit for the Chevrolet and several other cars, and like the Ford-Stromberg carbs, has been found to work very well on souped engines. It is a single-barrel type (see Fig. 6-3) featuring a removable venturi; diameters are available from 1-1/32 to 1-7/32 in. The main jet can be changed in ten minutes and a very wide selection of sizes are available from 0.040 to 0.078 in. (stock is 0.057). Internal flow losses are low.

*Zenith "28" Series* — This is a more expensive truck carburetor which is often used on souped Chevys. It is also a single-barrel type with the removable venturi (see Fig. 6-4). The feature that endears this carb to the hearts of competition men, however, is the very wide range available of metering parts — main, power, idle, and compensating jets, air bleeds, needles, etc. There are also a number of venturi sizes supplied, in increments of 1 millimeter from about 1⅛ to 1⅜ in. The Zenith is also a fine design from the standpoint of flow losses.

*Rochester "B"* — This carb is manufactured by a G.M. division and is standard on the late Chevys. It comes in two sizes (single-barrel) — 1.22 in. for the standard engine and 1.34 in. for the Powerglide. This carb uses a

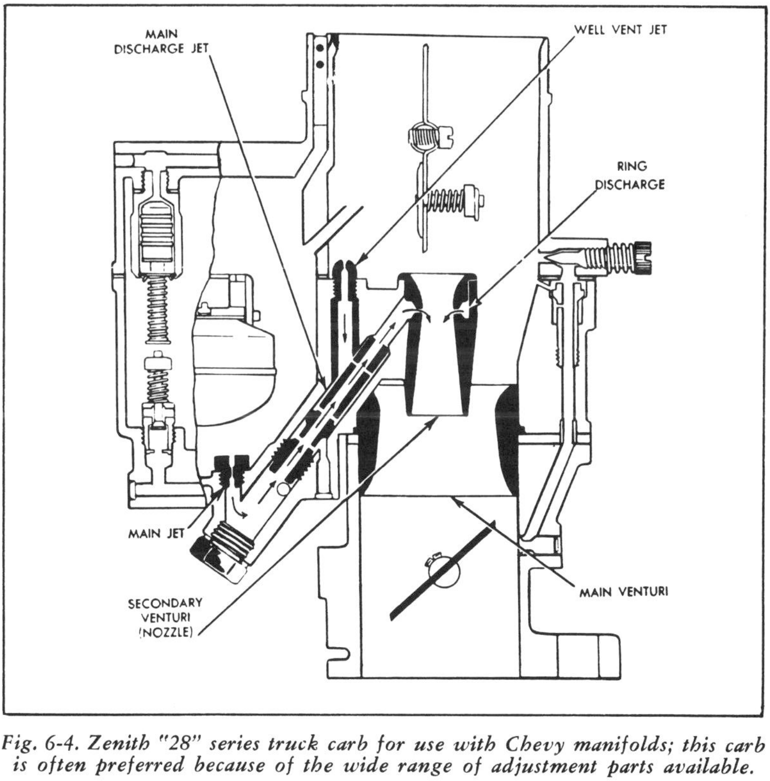


*Fig. 6-4. Zenith "28" series truck carb for use with Chevy manifolds; this carb is often preferred because of the wide range of adjustment parts available.*

conventional main jet (available in one lean and one rich size), but the nozzle bar across the venturi increases flow losses somewhat at high speed. The Rochester is frequently used for souped Chevy road engines, but not often in racing.

(Speaking of racing, a lot of fellows prefer to use the old dual-throat Ford-Stromberg "48" or "97" carbs because of less flow losses; special mounting pads are available to adapt these carbs to the standard Chevy manifolds.)

## CARBURETION TUNING

Be sure the throttle linkage is synchronized to give exactly equal butterfly opening (follow the manifold manufacturer's instructions carefully on this). Synchronizing the idling is quite simple: Hook a vacuum gauge line into the windshield wiper connection on the manifold and adjust the carb idle screws until maximum vacuum is obtained (which should be 15-20 in. with dual carbs). If you have no vacuum gauge, a rubber tube can be used like a stethoscope against the carb barrel; adjust till you get the same sucking sound from both carbs.

Now a word about jets. Getting the "right" jets in any souped engine is pretty much a matter of experiment, and even so, it's hard to tell anything definite. The best way of telling is to clock top speed with various jet sets. This may not always be possible, and in this case, you'll be pretty close if you select jets that will give the best acceleration through the gears. (In this type of experiment, clock over a measured course, use a slow rolling start, and shift at a pre-determined RPM or MPH.) An alternative procedure is to accelerate fast up to about 75 mph, declutch, shut the engine off, and then check the spark plug tips — if they are wet and sooty, the jets are too rich for maximum power; if burned and glazed, they are too lean.

Obviously all this will be a lot of work. And it may not add one HP to your output. But how can you be sure? We suggest that the serious souper who is building up a really hot engine should dig in and run full carburetion tests.

If you don't want to get into this, you can follow the manifold manufacturer's recommendations on jets. They usually take advantage of the improved mixture distribution with their compound manifolds and specify jets a step or two lean (from standard) to give better gas mileage. However, in several cases this has caused a considerable loss in top speed and performance compared with standard jets; experiment if you wish, but we recommend that you go "rich" if you are not satisfied.

For competition, and fuels other than gasoline, the fuel-air mixture ratio for maximum power is apt to be a lot different and we will need larger or smaller jet sizes under similar conditions. Below is a list of relative jet sizes needed for the various fuels under otherwise equal conditions, and based on gasoline as "100":

Gasoline — 100
Benzol — 90
Ethanol — 151
Methanol — 195
Acetone — 142
Ether — 157

For example, a 50-50 mixture of benzol and methanol would have an average index of 142½, so if our jet size with gasoline were 0.060 in., then we would need 1.425 x 0.060 = 0.086 in. jets for the new fuel. (If this size is not available, you can drill out a smaller jet.) Also remember the special conversion kits on the market for converting several stock carb models for pure methanol; these are a good investment if you are going into racing.

## FUEL INJECTION

Here is a fantastic and wonderful souping trick that has come into great popularity in the last couple of years for stock engines as well as the special Offenhauser racing types. The idea is to spray the fual directly into the air as it enters the port. Invented by Stuart Hilborn shortly after the war, complete units for the Chevrolet are now supplied by two companies; *Hilborn-Travers* builds one for the Wayne head, and *Howard's* has a unit to fit the 3-port stock Chevy head.

Fig. 6-5 shows the H-T unit. A pump feeds the fuel under pressure to the spray nozzles in the port section; air flow is controlled by conventional butterfly throttles, which are linked up with the pump to vary the fuel feed with the air flow. This carburetion setup boosts the HP output for three reasons: (1) Better breathing because of no flow restriction in manifold and carbs, (2) absolutely equal mixture distribution to all cylinders, and (3) greater charge density because of no heat transfer to the mixture in the manifold. Tests with this equipment by Hilborn-Travers have shown HP increases over conventional carburetion of never less than 9%, and some increases went to 34%!

This fuel injection idea is beautiful for racing. Because of improved mixture distribution, fuel mileage increases of 50% are common, which is always good in racing. Air filtering was a problem at first, but improved filter heads are licking this (though you seldom see fuel injection on dirt tracks). Costs are quite high, however, the H-T equipment running about $175 for the Chevy, but we strongly recommend you consider it carefully for racing, especially on semi-paved tracks. Incidentally, this carburetion setup is not practical for road engines.

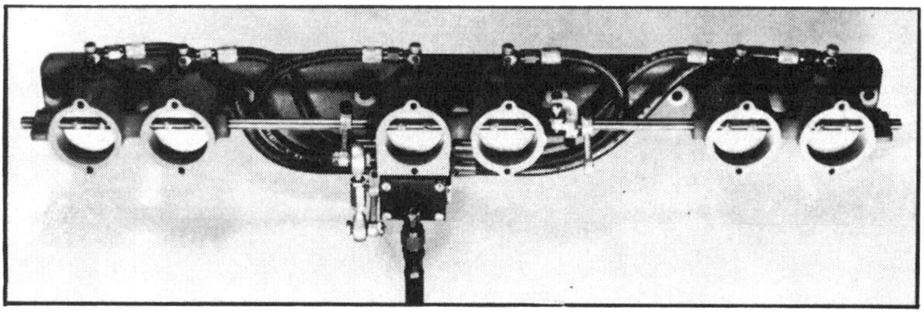

*Fig. 6-5. Hilborn-Travers fuel injection unit to fit the 6-port Wayne-Chevrolet head. Note how the throttle valves are linked with the fuel distribution block to vary the fuel feed with the air flow; approximately 1.0 hp/cu. in. can be obtained using this equipment with a full-race setup.*

## MANIFOLDS

As we learned in the last chapter, one way to reduce induction system pressure losses is to reduce the flow rate or velocity. This can be easily accomplished by using multiple carbs and manifolding. With dual carburetion, we cut flow losses between carb and port by about 75%, and this will raise peak HP some 18% on the Chevy (see Fig. 6-6). In fact, dual carburetion eliminates practically all induction losses upstream of the port, so that adding a third carb would only boost HP another 3-5% at the peak! (This shows conclusively that the great majority of the flow losses come between the head port opening and the cylinder, which emphasizes the importance of porting and valve modifications.)

However, there is another more obscure factor in this flow problem which can make quite a difference, and that is the manifold *contour* or shape. With the straight 90° type of manifold used on 6-cylinder engines,

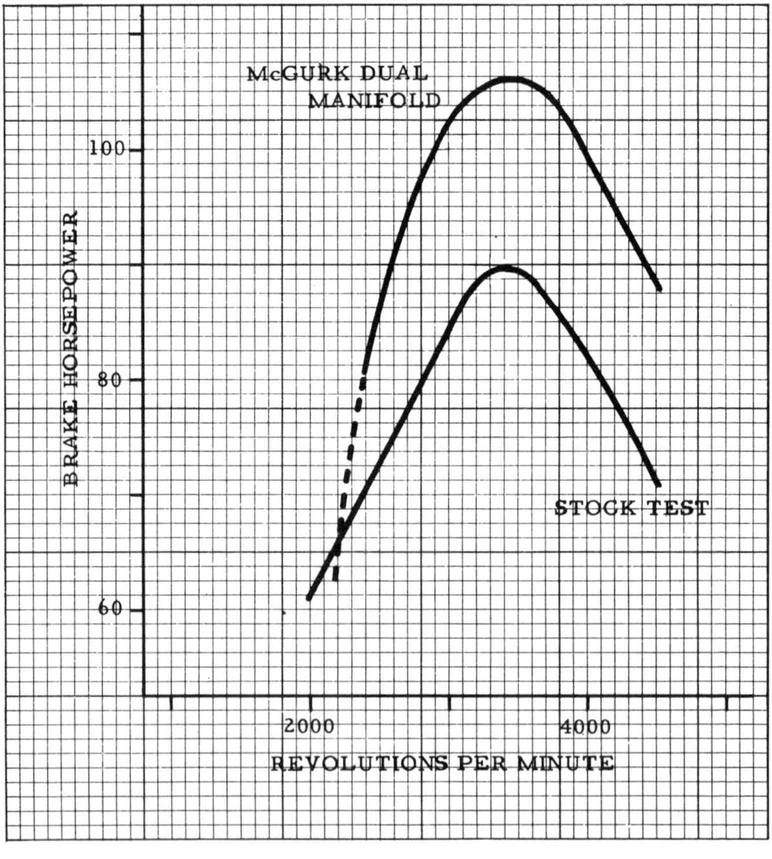


*Fig. 6-6. Dynamometer test of a McGurk dual manifold on the stock Chevrolet engine; no other changes (running without accessories).*

it is vital that we have smooth bends of large radius to give the minimum losses through turbulence in the fuel-air mixture. Fig. 6-7 shows what we mean; Manifold A will have at least twice the flow loss of Manifold B — and laboratory tests show that just one sharp 90° "L" will lose 15 times as much pressure as all the surface skin friction in a manifold. So we say: For maximum HP, go for the manifolds with the wide bends.

Another important point about performance with multiple carburetion: You get practically no boost in HP below 2000 rpm — in fact, you will probably get a slight loss, assuming no other changes. This is because, with more than one carburetor, gas velocity in the manifold at low RPM is so low that gas *momentum* is practically non-existent. Thus with the late closing of the inlet valves, a condition will be reached where the pistons coming up on the compression stroke will *reverse* the flow in the manifold

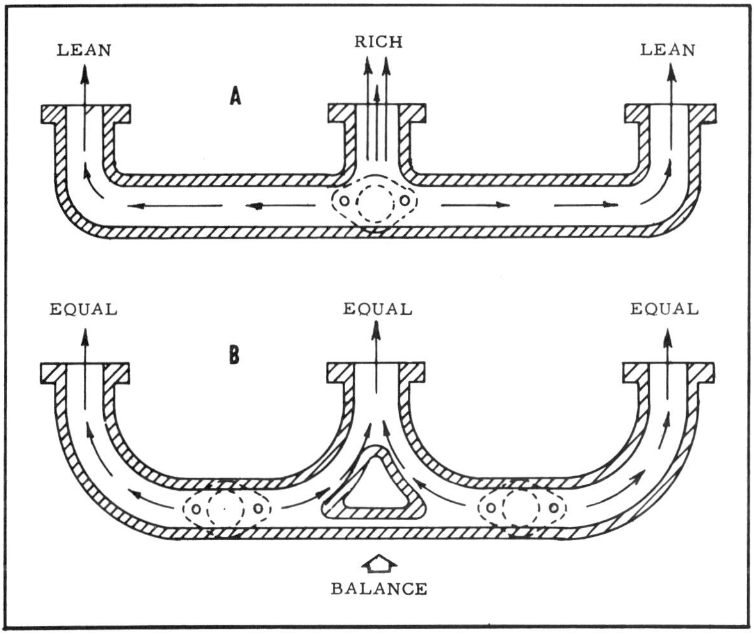


*Fig. 6-7. Drawings illustrating the importance of wide-radius bends in reducing flow losses in a manifold; Manifold "B" is the preferred setup.*

and play hob with mixture distribution. So don't stick dual carbs on your engine and expect big things at low speed. And also don't expect any great improvements with three carbs instead of two.

Getting down to practical cases, we find the manifold situation well taken care of in the Chevy souping field. There is a wide choice of dual and triple units now available from such outfits as McGurk, Nicson, Tattersfield, Edelbrock, Edmunds, Weiand, etc. These range from about $25

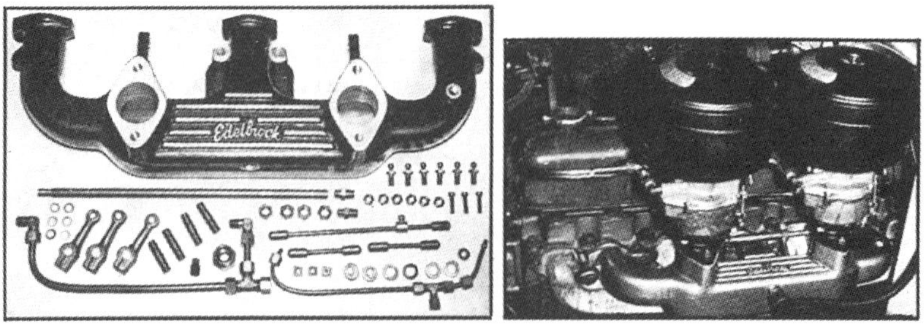


*Fig. 6-8. Edelbrock dual exhaust-heated manifold set for the Chevy.*

*Fig. 6-9. Tattersfield dual exhaust-heated road manifold; note the smooth sweeping bends.*

*Fig. 6-10. Edmunds dual manifold setup for the Chevy with water heating (exhaust riser blocked); this layout gives more steady, even heating.*

*Fig. 6-11. Newhouse dual Chevy road manifold; note throttle linkage.*

to \$45 for the complete unit, including throttle linkage. Some of the units are shown in accompanying illustrations.

The manifolds designed for road engines have either water or exhaust-heated "hot spots" which can be run open or blocked, as desired. This manifold heat is desirable for a road engine to vaporize the fuel, give better mixture distribution and more flexible performance in city driving. However, the heat also expands the charge, which means less weight drawn in and less HP (it can make an easy difference of 10%). We recommend thermostatically-controlled manifold heat for all road engines except perhaps a special job for fast highway travel; naturally hot spots are *out* for racing.

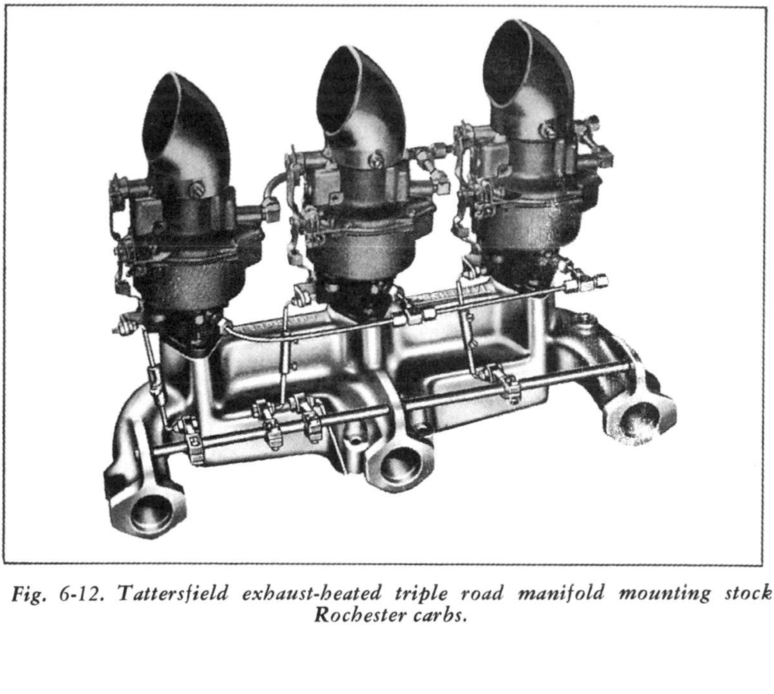

*Fig. 6-12. Tattersfield exhaust-heated triple road manifold mounting stock Rochester carbs.*

*Fig. 6-13. Above—Nicson triple road manifold with Zenith carbs.*

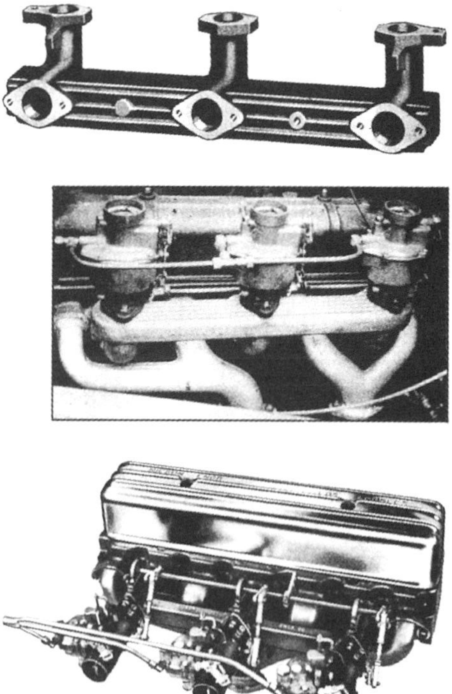

*Fig. 6-14. Left—three special Chevrolet triple manifolds for racing, without hot spot—top, McGurk; center, Lee; bottom, Nicson, for Zenith 8810 side-draft carbs.*

*Fig. 6-15. Special multiple manifold for the Wayne head mounting six Strom-Berg BXOV-2 carbs; this setup is used on Marv Lee's straightaway "Lakes" streamliner( not practical for road or track).*

## VALVES

Larger than stock exhaust valves have no appreciable effect on HP and are usually left at stock size (1.50 in.). However, the stock exhaust valves lose a lot of their hardness at very high temperatures and are apt to pit and burn excessively under conditions of sustained high output. The heavy-duty nickel steel exhaust valves for the GMC "270" and "302" truck engines will give much better service and can be used without any changes in both the standard and Powerglide heads. We recommend these for the hotter souping jobs.

As for intake valves, diameters larger than stock are becoming quite popular. On the standard head, stock diameter is 1.64 in., and you can fit up to 1⅞ in. (1.88 in.) valves on the late heads—slightly larger on pre-1941s. The Powerglide engine carries a 1.94-in. intake valve; this is very large (second largest of all U. S. stock cars) and there seems little point in trying to fit a larger one, since it could be only slightly larger anyway. In this case, just open up the port beneath the valve.

Now what can we get in the line of replacement valves? There's not much of a choice here because most stock engines use 45° intake seats. Powerglide valves can be fitted in early heads, but casting sections are a bit thin for this on '41 and later heads; for these heads, you can use the 1.86-in. GMC "270" valve with some stem modifications. Several companies supply special oversize valves to fit the Chevy in diameters around

1⅞ in. and at prices in the neighborhood of $10 a set. These are probably the most convenient.

Here are a few tips on fitting oversize intake valves: Bore out the original valve port an amount sufficient to give a seat width of about 1/16 in. near the upper edge of the seat taper, then face off a new seat in the head at 30° (this seat width is sometimes made smaller on racing jobs). It's also a good idea to open out the cylinder head around the valve at least ¼ in. larger than the valve head diameter to prevent the incoming gas flow from being "masked". You can bore the port width ¾ in. or so into the head, from which point you can taper it into the original port line with a grinder. Incidentally, don't try cutting off the valve guides where they extend into the port; this will cut flow losses a bit, but you can't keep guides in with this shortened length.

## VALVE SPRINGS

The stock Chevy valve spring is a very durable piece of equipment, and will work out well under high-output conditions. However, when we use a special camshaft with faster opening and closing rates and more dwell, we find that the spring tension is not sufficient to keep the tappet on the cam at high RPM. This is known as valve "floating". The first result is a loss of power, and as the RPM is increased further, the engine begins to miss and finally cut out; also this floating literally pounds the valve gear to pieces. On the Chevy engine, the valve floating problem is especially acute as compared with an L-head engine because of the extra weight of pushrods and rocker arms — which must be accelerated by the spring tension to follow the cam on the closing side.

On late engines, tension of the stock springs with the valve open is 132 lbs. on the standard and 160 lbs. on the Powerglide (both are interchangeable). With a ¾-reground cam, the stock setup would likely start to float badly at around 4000 rpm, depending on a number of factors. A well-reground cam should have at least 180 lbs. tension on the standard and 200 with the heavier Powerglide valves.

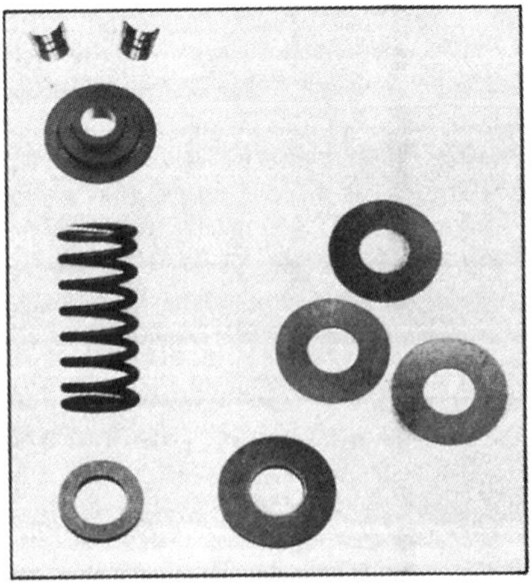

*Fig. 6-16. Wayne inner valve spring kit, including washers and shims to permit tension adjustment, and a special heavy-duty spring retainer to insure against "swallowing" a valve.*

There are several choices of stock spring replacements available. We can use the stock Buick inner and outer spring set without changes for a total open tension of 170 lbs.; or we can use just the Buick inner spring with the Chevy spring to add about 50 lbs. to the open tension. Several companies, including Wayne, supply special inner springs to use in conjunction with the stock Chevy spring; these boost total tension to around 250 lbs. and are nearly a "must" for the high-RPM track engine (See Fig. 6-16). At any rate, figure on some spring changes with any reground cam.

When using dual springs, always have the coils winding in *opposite* directions to prevent "surge". The stock Chevy spring retainers are somewhat flimsy; you can easily "swallow" a valve at high RPM with a radical cam; special retainers are available and these are recommended for racing engines. When using Chevy springs with various head setups, be sure the length is close to the stock 1-27/32 in. with valve closed, to assure proper tension (you may have to either shim it up or countersink).

## ROCKER ARMS, PUSHRODS, TAPPETS

The stock Chevy "valve train" has been found quite adequate for high outputs, so extensive changes here are not necessary. Special intake rocker arms are available at around $12 a set with long valve lever arms to increase lift 1/16 in. over stock (see Fig. 6-17). This increase in valve discharge area will add 3 or 4 hp at the top, but we doubt much effect at low speed.

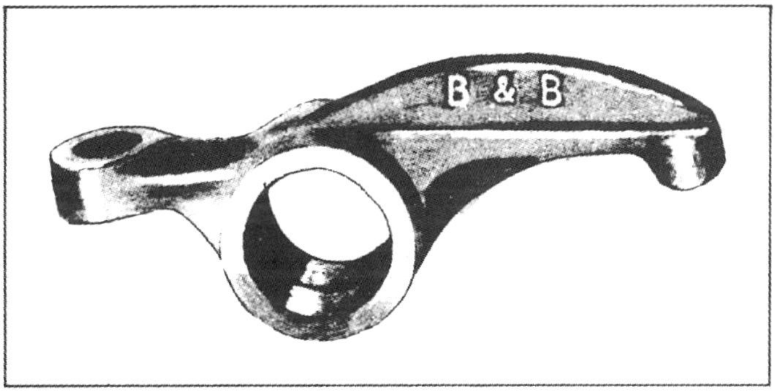


***Fig. 6-17. Barker high-lift rocker arm; it has long lever arm to the valve side and gives 1/16 in. higher lift.***

The stock Chevy solid pushrods will be fine for most conditions. However, with fierce cam shapes and sustained high-RPM as in track racing, the stock pushrods have a tendency to flex or bend. Tubular rods will solve this; you can use either stock Buick rods in conjunction with a 1937 Chev cup-type tappet (with some tip modifications on the rod), or one of the special tubular pushrod sets now on the market (see Fig. 6-18). For most cases, the regular Chevy pushrods should be satisfactory.

Similarly with tappets, the early "barrel" types work out very well under all conditions. The late hydraulic tappets on the Powerglide, how-

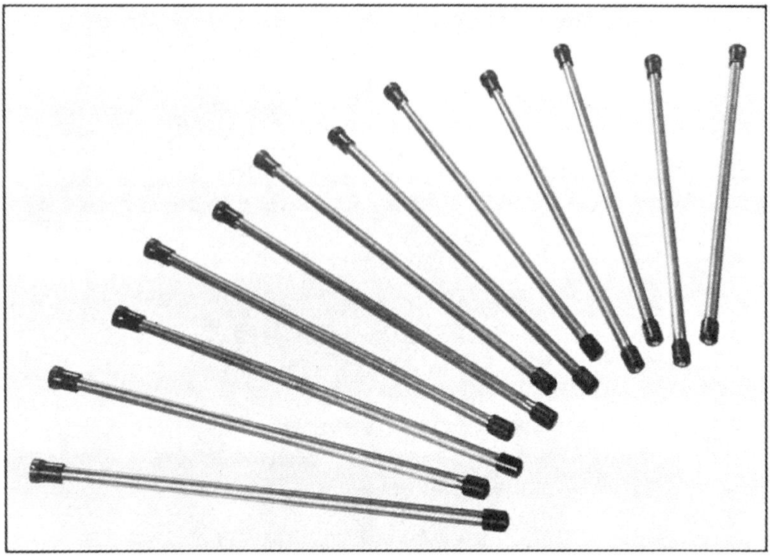

*Fig. 6-18. Wayne tubular pushrod set.*

ever, are not good for high RPM. An inherent characteristic of this type of tappet is that it will start to "pump up" as soon as there is any slack in the valve train, such as would be caused by mild valve float. This, in turn, holds the valve off its seat on the return stroke, and the engine starts to miss. We have no exact figures, but it is unlikely that you'd be able to turn a reground cam much above 3500 rpm with hydraulic tappets. Naturally this is unwanted. You can quickly replace these units with stock Chevy Hi-Torque truck tappets without changes.

## THE CAMSHAFT

Here is one of Leadfoot Louie's favorite stamping grounds. He has gone all the way from "regrinding" his cam with a file to trying to lathe one up from a solid billet! He never did get one to work right, though. The nearest he came was getting one to *run*. But it was rough, and the higher he revved it to smooth it out, the rougher it seemed to get – it was still sputtering when No. 2 rod came through at 6500!

Cam reworking is no job for an amateur. There are a lot of shops around, especially in California, that specialize in this work at reasonable prices, and it's best to utilize their services in this case.

What do we mean by cam "reworking?" In Chapter 3 we stated that you can increase the weight of fuel-air charge inducted on the intake stroke at high RPM by starting to open the intake valve before top center and leaving it open well after bottom center – thereby utilizing the momentum of the incoming gas flow to pack more charge into the cylinder while the piston is coming up. The longer this valve open period, the better the volumetric efficiency at high speeds.

At the same time, as we lengthen the period, the peak of our V.E.

curve moves to higher and higher speeds (see Fig. 3-1, Chapter 3) so that the charge — and HP — fall off drastically at low speeds. The inertia which is pushing the gas at high velocity just isn't available at low speed, and the flow is reversing out the valve as the piston comes up. This can get so serious with some timings that the engine will barely idle at 1000 rpm and won't pull noticeably in the usual gear ratios below 35 mph! These facts are the basis of our cam problem. We can rework for good power output at high RPM, but this will kill off performance at low speed where we need it for city driving. If we want both, we must compromise.

Cam regrinding is well-known. In this practice, possibly 0.100 in. is ground off from the heel of the cam and the sides are rounded up; this lengthens the valve open period, increases the *rate* of opening and closing, and increases the lift. Fig. 6-19 shows what is done. There are a large number of shops around the country that will regrind your stock Chevy cam to any specification for generally \$30-\$40.

Now by varying the regrinding contour, we can get about any combination of period, rate, and lift we want (within the limits of the original

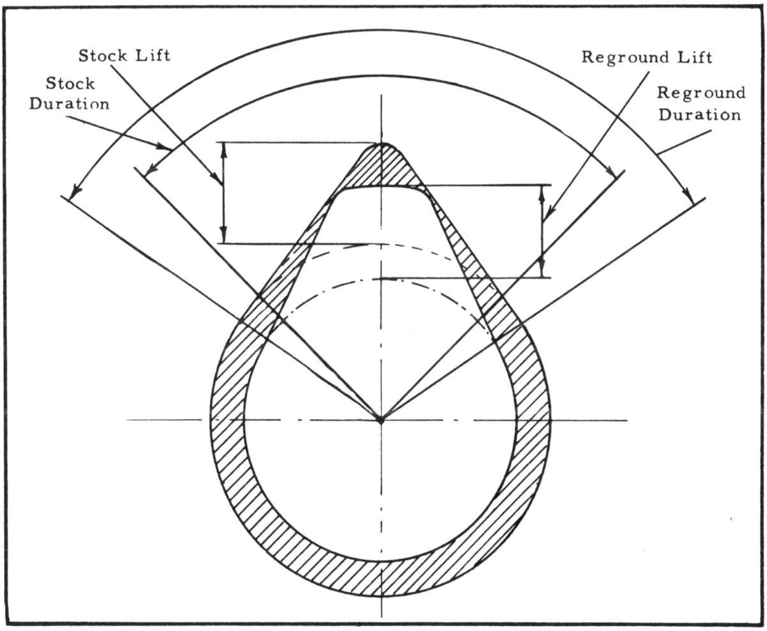


*Fig. 6-19. Drawing showing the stock taken off when regrinding a cam.*

cam outline), to fit any performance requirements. Since the stock Chevy cam timing is relatively mild, regrinding will do wonders here. Fig. 6-20 shows the approximate HP curves obtainable with two degrees of cam grinding, primarily to illustrate the effect on low-speed output. Don't forget this vital item!

In order to eliminate the confusion that cropped up several years ago over the various "grinds", the grinding companies have more or less agreed on a standard set of designations for the different degrees of grinding, based on an air flow feed equivalent to stock piston displacement (216 cu. in.). Here is a run-down:

*Semi Grind* – This is the mildest grind, timed about 15-40 on the intake (degrees opening before and closing after dead center) and 50-10 exhaust.

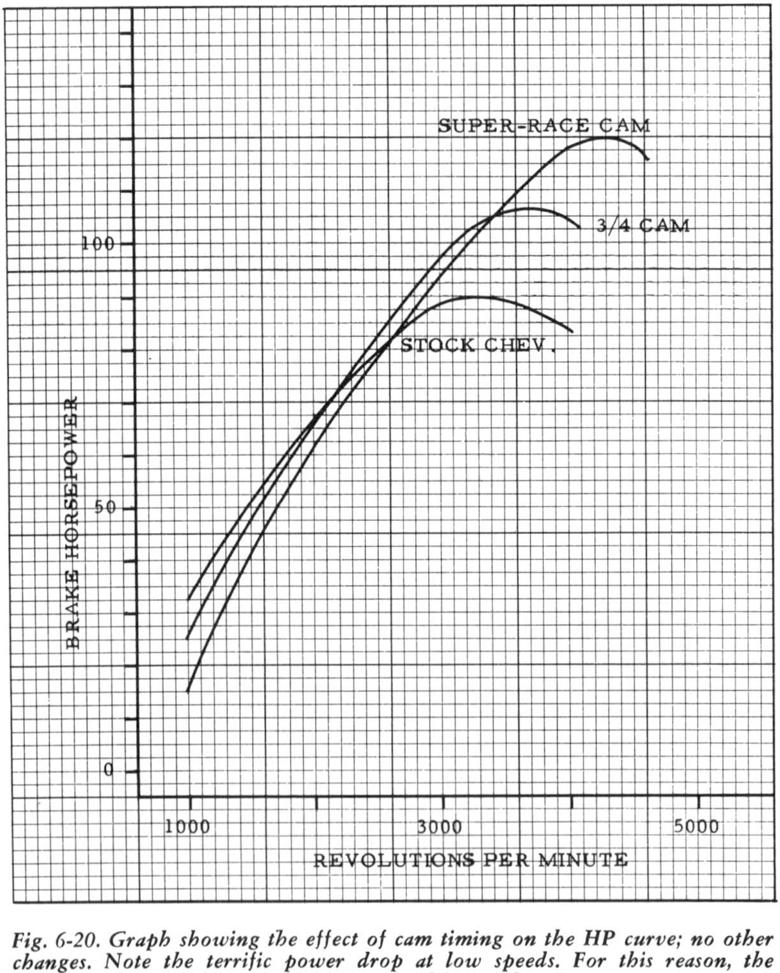


*Fig. 6-20. Graph showing the effect of cam timing on the HP curve; no other changes. Note the terrific power drop at low speeds. For this reason, the "super" cams are not good for road engines.*

This one will boost peak HP about 10% over stock, but it's too mild to be of much practical use in souping.

*Three-Quarter Grind* — This is a good one for road work with stock displacement, timed about 25-50 intake and 55-15 exhaust. It raises HP 15-20% over stock and still gives fairly good lugging at low speeds (see Fig. 6-21).

*Full-Race Grind* — This one is timed about 30-65 intake and 65-25 exhaust. It will boost HP at least 20% over stock and is quite suitable for a road engine with over-stock displacement (around 230 cu. in. or more), though it might run fairly rough at low speed with 216 cu. in. Even then, however, output at 1000 rpm with a "full house" will be less than stock.

*Super-Race Grind* — This is strictly a track grind, timed about 35-75 intake

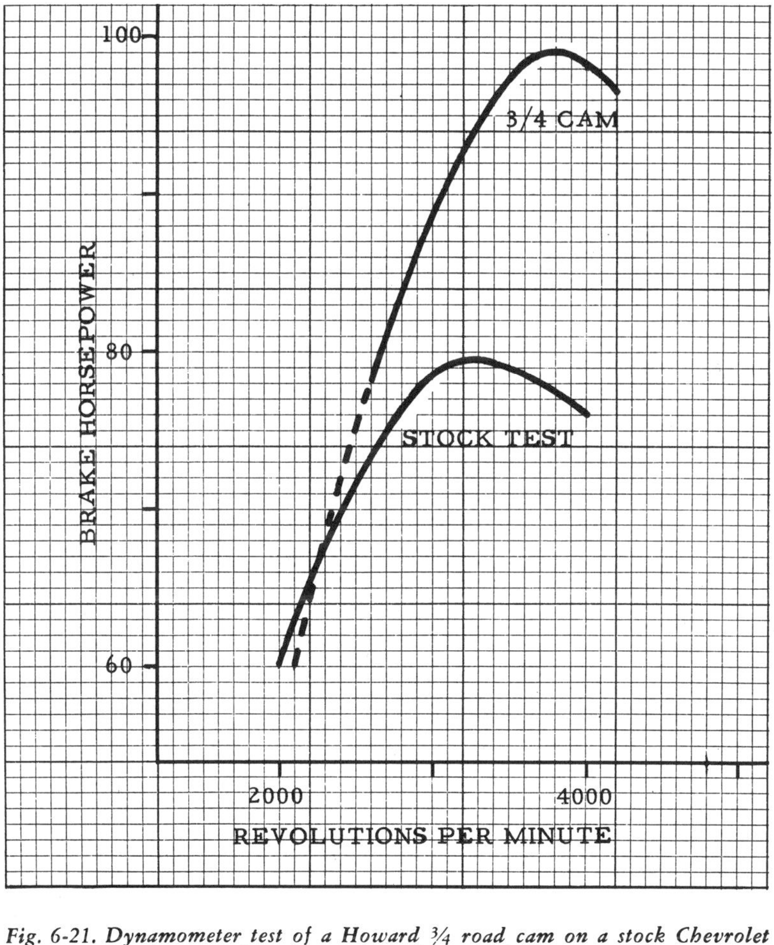


*Fig. 6-21. Dynamometer test of a Howard ¾ road cam on a stock Chevrolet engine; no other changes (running with full accessories).*

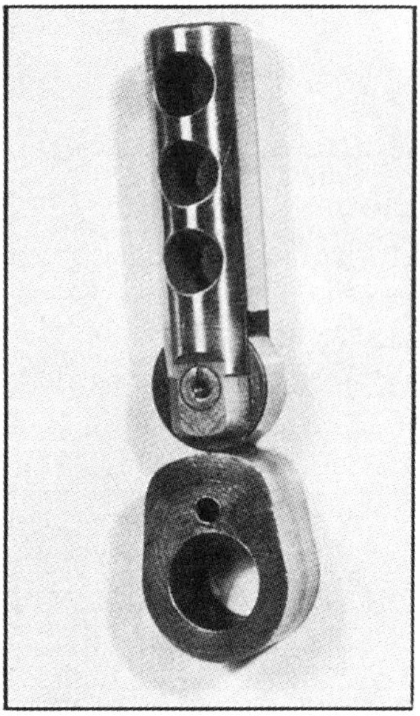

*Fig. 6-22. Chet Herbert's roller cam and tappet setup for the Chevy and GMC; this track cam has 290° of intake duration, and note the very quick opening and long dwell.*

and 75-30 exhaust, and giving a peak power boost of 25-30% due to the cam alone. Peak torque comes at around 3000 rpm, however, so don't expect much of the Super grind in street traffic (you might get away with it at 250 cu. in., but we won't advise it for a road engine).

From this point each grinding company usually has its own pet "super-grinder" grind for all-out competition work, which gives a few more HP at the top and less and less at the bottom. These shouldn't be tried for road use under any conditions. As a matter of fact, stock cam regrinding is about at its limit. This is because of the narrow tappet base; with more and more metal ground off the cam heel, the toe becomes relatively longer in relation to the cam "base circle" — until a point is reached where the toe will strike the edge of the tappet coming around, rather than sliding under it.

So what is a solution? One answer is the new "roller-tappet" cam sets (see Fig. 6-22). These replace the stock wide-base tappet with a special light unit having a *roller* cam follower and running in a sleeve in the stock tappet boss. The effect is to give an *infinite* tappet base diameter, which allows any combination of period, rate, or lift beyond what we can get with cam regrinding. Special cams must be used, of course. Right now they are being run up to nearly 300° of intake duration, and the new timing and lifts have shown a HP margin of some 5-7% over regrinds so far. The setups are expensive, though, so they need only be considered for super competition engines.

Before we leave the subject of cams, may we repeat: Cam regrinding is a very economical and effective method of boosting peak HP, but it's a killer at low speed. You can stick on all the carbs, ignitions, and heads in the world, and get a terrific power boost over the full speed range — but when you add that cam, it will make the whole unit look sick below 2000 rpm.

## THE EXHAUST SYSTEM

Here is a little matter that's often forgotten by the amateur souper in his enthusiasm for cams, carbs, and heads. Actually there is no easier way to get another 8 or 10 hp than to rip off that stock muffler and tailpipe.

As we all know, the stock exhaust system is hard on HP at high RPM. This equipment has been designed with an eye both to cost and silent running – not HP. As a result, the gas flow restriction through the single pipe and baffle-type muffler at high speeds will mount to some 3-4 lbs./sq. in. "back-pressure" on the piston head. This costs precious HP and gas mileage.

There are several things we can do to relieve this condition. For racing, of course, we can run individual headers from each exhaust port to a large open pipe. For the road, however, John Law would hardly smile on this arrangement and we must make at least some pretense at muffling. The most simple move here is to replace the stock muffler with one of the new "Hollywood" straight-through types. These carry the exhaust gas straight through the muffler by way of a perforated tube packed in steel wool; they merely deaden the sound, and do not actually muffle it (see Fig. 6-23). A setup like this will reduce back-pressure to around 1 lb.

The next logical step in combating exhaust losses is to cut the flow rate in half by using two pipe lines. Several companies sell dual systems for the Chevy to fit on the stock exhaust manifold; using stock mufflers,

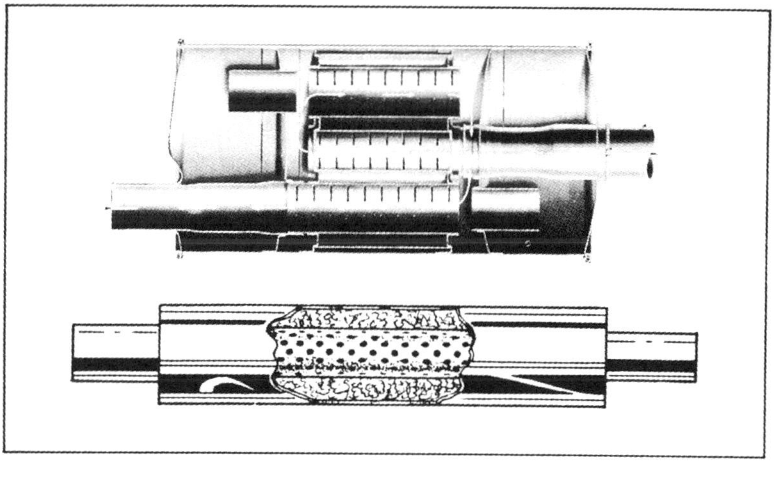

*Fig. 6-23. Constructional differences between the stock Chevy muffler (top) and a "Hollywood" straight-through type. The stock muffler baffles the gas flow, whereas the straight-through merely deadens the sound without obstructing the flow in any way.*

*Fig. 6-24. Clark dual-exhaust headers for the Chevrolet.*

back-pressure is cut to ½ lb., and to ¼ lb. with Hollywoods. These sets sell for around $30.

The ultimate in exhaust gas flow refinement is obtained by using the Clark racing-type double exhaust headers in conjunction with dual lines and straight-through mufflers (see Fig. 6-24). Complete sets in this class sell at around $70, and tests show no appreciable back-pressure up to peak RPM.

Incidentally, don't forget that the exhaust noise level is going up while you cut the back-pressure. Horsepower comes with a price tag on it, no matter where you get it. So if you long for the quiet life, don't go for dual Hollywoods! However, a dual exhaust system is a "must" for any souping job that runs the HP more than 120 hp.

In conclusion, let us restate — and redigest — that one big factor in all this work: Induction and exhaust refinements are *effective* only at *high* engine speeds, above 2500 rpm; if you want "guts" at low speed, you are wasting your time in this field. And if you try to get it with a reground cam — you are "stabbing yourself in the back!" (If you ever felt a full-race cam "come in" at 55 mph, you'd know what we mean.)

Our souping job has just begun. We have our charge into the cylinder and a lower end that will put the motion to the flywheel. The next thing is to fire it at just the right instant; there is more to this than meets the eye.

# CHAPTER 7

## IGNITION

TO GET the maximum torque on the flywheel, we must start combustion in each cylinder at precisely the right instant on the compression stroke — the "right instant" in this case, varying considerably with load and RPM.

That, in essence, is why we're devoting an entire chapter to an item that doesn't pack a HP in a carload. Ignition is too often the forgotten factor in the amateur souper's "horsepower pipe dreams". Because the manufacturers of special ignition equipment can't promise a flock of extra horses, Leadfoot Louies won't bother to think about this in their plans.

But we can't help but remember the souper who dropped a full-race Chevy in his "225" hydroplane. He tells us, "I've got a full house; 12:1 compression, triple carbs, ported, super cam, GMC rods, stock ignition... Wonder what's wrong with it? Starts to cut out around 4000. Must be carburetion." It wasn't carburetion. It was ignition; and he'll never get it up to peak RPM under full throttle until he makes a few changes in this department.

So let's pay some close attention to this vital factor in our souping plans, and unlock the door to those horses we have in the Chevy block.

### BASIC FUNDAMENTALS

Before we can intelligently approach the problems involved in ignitioning the souped engine, we must first understand the basic ignition requirements and the limitations of the original stock layout.

And it will help if we pause here and get one thing firmly in mind: Electric current acts very much like a liquid or gas flowing in a closed system. In other words, we have to work with concepts such as pressure, friction, momentum, etc. — only with electric current, we use the terms: volts, amperes, line drop, impedance. But the idea is the same. When electric current flows through a wire, it requires a pressure or voltage to move it; the "friction" of the flow appears as a voltage or pressure drop. Similarly, electric current has a property something like *weight*; it possesses a definite inertia, and you can't start it or stop it instantly. Keep these facts always in mind when you study ignition.

We won't go into a lot of detail here on ignition fundamentals because we're sure the readers know how a battery-coil system works. We need only look at the basic problem now — which is, of course, to get a good hot spark, with no lag, under all conditions and at all RPM. There's more to this than meets the eye.

In the first place, we have two strikes on us from the start because, in all usual souping work, we are actually plotting against the ignition system! Combustion theory teaches that the current voltage required to jump a given spark gap is proportional to the gas density between the electrodes. Therefore, when we port, install large valves and hot cams, and raise compression ratio, we just boost that compression pressure and increase the voltage required to throw a spark. This is worse than it sounds. With a

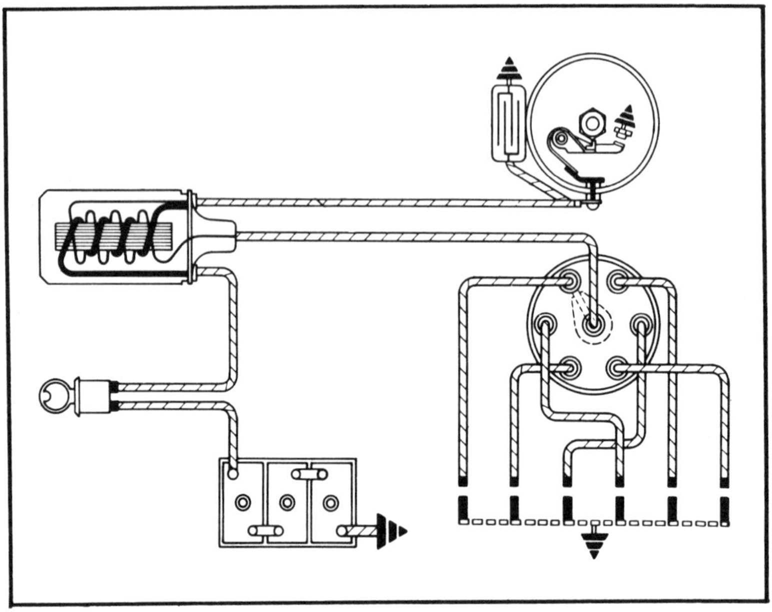

*Fig. 7-1. Wiring diagram for a typical 6-cylinder battery-coil stock ignition system.*

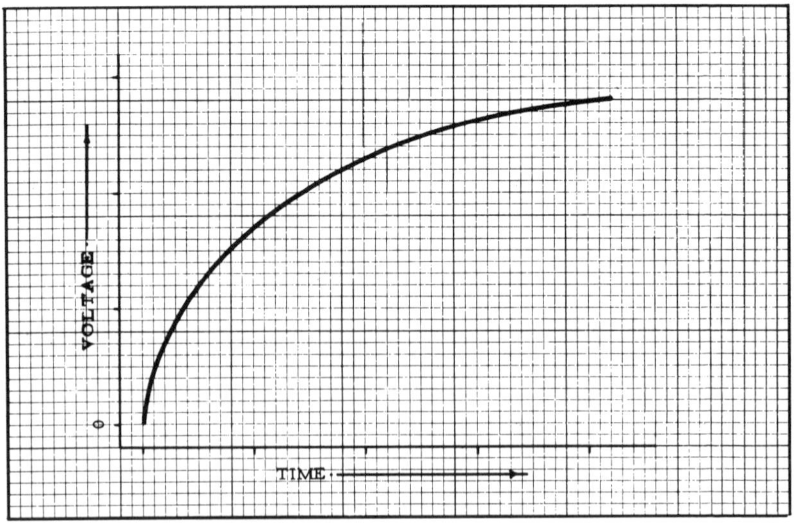


*Fig. 7-2 Generalized curve showing how high voltage builds up in the primary coil when the breaker points are closed.*

full house at 9:1 compression, for instance, spark voltage required at a given RPM with full throttle would be over 50% higher than stock. That's a big boost to ask stock ignition to handle.

And this is only half the problem. There is a basic limitation of the battery-coil ignition system that is working against us too — the fact that the available spark voltage drops off fast as RPM goes up. Here's why: Electric current has "inertia," and we can't just close a circuit and have it start and reach full flow instantaneously. Therefore, when the breaker points close, and start the current flowing from battery to coil, the voltage builds up gradually in the primary winding — as in Fig. 7-2.

With a given feed pressure of 6 volts from the battery, the time required to build up full primary voltage is virtually independent of RPM. So we find that as speed increases, the points are closed a shorter and shorter time, and the primary current is being chopped off before it has time to get going. This, of course, reduces the voltage output of the secondary coil to the spark plugs in proportion. Fig. 7-3 shows how spark voltage drops with RPM for a typical stock car installation.

Stick this same layout on a souped engine with 8 or 10:1 compression, wind it up over 4000 rpm, and what happens? The spark voltage is not

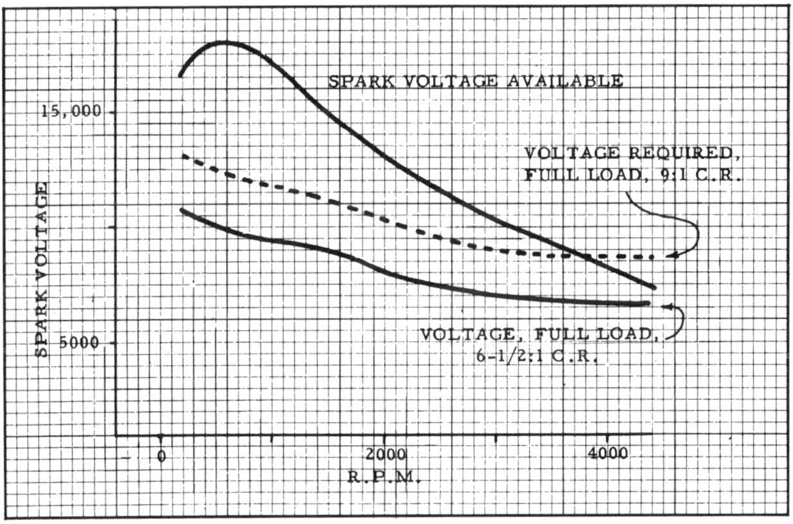


*Fig. 7-3. How spark voltage drops off with RPM on a conventional single-breaker ignition, because of less time available for saturating the primary coil; the lower curves show voltage required to throw a spark.*

sufficient to jump the gap against the high cylinder pressures, and the engine sputters and cuts out. It's hard to say just where a certain stock ignition will fail with such-and-such a compression ratio with a certain fuel, but it is certain that something will have to be done on anything above a moderate souping job.

Just what can we do? If we consider only reworking stock ignition, two major possibilities appear: (1) Use a special high-output coil that will give a peak spark pressure of around 30,000 volts instead of the usual 20,000 for stock, and (2) use two separate breaker-coil systems with half the cylinders on one and half on the other, so that we can double the time the breakers are closed at a given RPM.

## REWORKING STOCK IGNITION

Chevrolet has always used a very rugged, straightforward Delco-Remy ignition system, with a conventional automatic vacuum-centrifugal spark advance mechanism (see Figs. 7-4 and 5). Actually, our ignition problem under souping conditions with the Chevy is not nearly so critical as that on the popular Ford-Merc V8 block. That's because we are working with only 6 cylinders instead of 8. The distributor cam "dwell" angle for saturating the coil (the degrees of distributor shaft rotation during which the breaker points remain closed) is 35° with the Chevy. The pre-1949

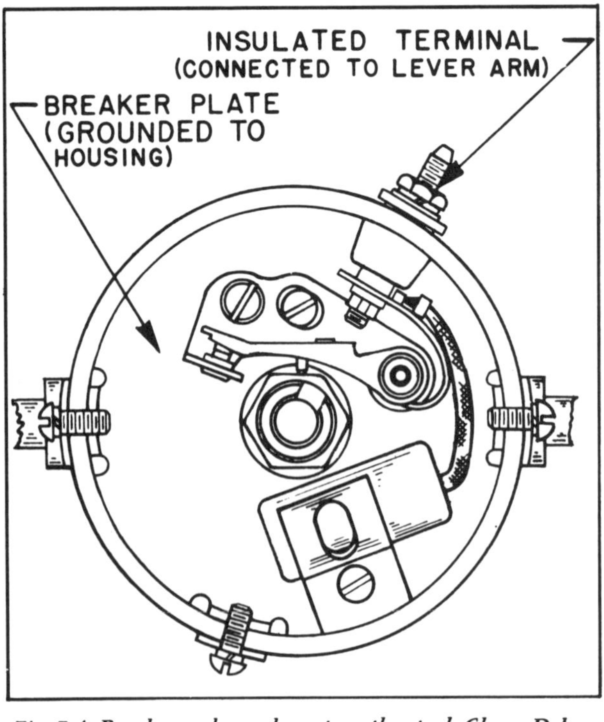


*Fig. 7-4. Breaker and cam layout on the stock Chevy Delco-Remy ignition.*

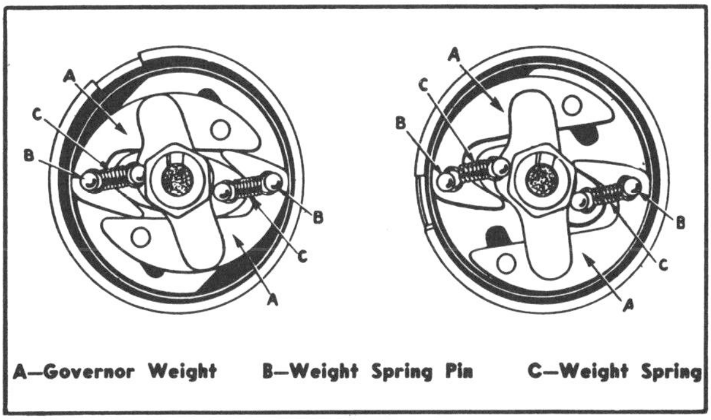


*Fig. 7-5. Centrifugal spark advance arrangement on the stock Chevy ignition.*

Ford equalled this with a clever double-breaker system, but the cam lobes had to be made quite sharp, and this invited breaker-arm floating. Consequently there are a number of conversion kits on the market for the Ford ignition with special arms, springs, etc. No such condition exists with the Chevy ignition; it will keep right on clicking along, after the V8 ignition has cut out.

So there's not much we need do with the stock system under mild souping conditions. For higher outputs, some "beefing up" is desirable, and a considerable amount of special equipment is now available for this purpose. The simplest step, of course, is to install one of the good, heavy-duty (30,000-volt) coils, such as are used on trucks. These are widely available from D.S.M., Mallory, Bosch "Big Brute", etc. at around $8-$12, and can be adapted very easily. These will take care of medium souping requirements at reasonable RPM.

The next step is a Mallory double-breaker conversion kit at about $7 (see Fig. 7-6). This setup works like the above-mentioned pre-'49 Ford

*Fig. 7-6. Mallory double-breaker ignition conversion kit for the Chevy; this can be installed by anyone in a few minutes.*

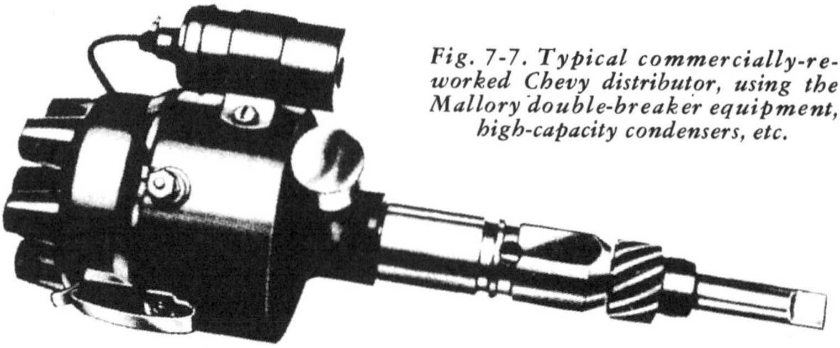

*Fig. 7-7. Typical commercially-reworked Chevy distributor, using the Mallory double-breaker equipment, high-capacity condensers, etc.*

ignition; two breakers are wired in parallel and positioned in relation to the breaker cam so that their "closed" periods *overlap* (since the circuit would be closed until both breakers are open). This little trick increases the effective cam dwell angle to 43° and is guaranteed to throw a spark up to 7000 rpm. The installation is very simple, and can be accomplished in fifteen minutes. Incidentally, several companies supply complete Mallory converted distributors on an exchange basis at prices in the neighborhood of $20. (See Fig. 7-7).

Now if you are thinking of trying to convert your stock ignition to a true dual breaker-coil layout, better forget it — unless you are a Clay Smith or some such. It's been done in backyard shops, but it's definitely no job for an amateur; there is just too much precision, skilled work necessary to get an ignition that will stand up. We suggest you shop around for a commercially-reworked unit, such as:

## CONVERTED DUAL SYSTEMS

The idea here is to use two separate breaker-coil circuits so that we have only three cylinders on one circuit and can double the time available

*Fig. 7-8. The Spalding dual breaker-coil conversion of the Chevy ignition using 3-lobe cam, two breaker circuits, dual coils, etc.*

for saturating the coil at a given RPM. We then use two equally-spaced breaker circuits, a 3-lobe cam, and get a cam dwell angle of around 90°.

As a matter of fact, there is not much available in this category for the Chevrolet. Spalding and several general equipment companies building up their own units comprise the field. All of them base their ignition conversions on the stock distributor (see Fig. 7-8) and guarantee a spark up to 10,000 rpm. Prices range generally under $50. These dual ignitions are completely satisfactory, from a performance standpoint, for the most severe souping conditions.

## MAGNETOS

The magneto doesn't hold the place it once did in the speed field. Time was when you couldn't do a thing with a hot engine without first sticking on a "mag" to handle the ignition problem. But today, with our highly-developed dual systems, a mag is only for the most extreme conditions.

A magneto is nothing more than a conventional breaker-coil ignition unit, except that the current is supplied by a built-in generator instead of a battery. It's all housed in one unit and can be conveniently driven off the stock distributor shaft. The major advantage of a mag is in its voltage output characteristics; since the armature cuts the magnetic field faster as the RPM increases, we have the delightful situation where our spark voltage actually increases with speed (though electrical inertia effects cause it to level off at high RPM). Fig. 7-9 shows an output curve for the Scintilla-Vertex mag. For this reason, and because of the compact, reliable

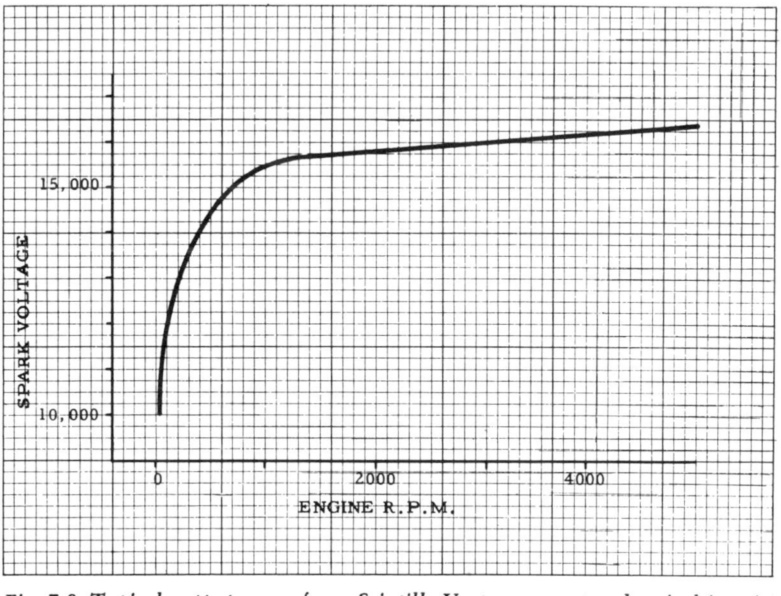


*Fig. 7-9. Typical output curve for a Scintilla-Vertex magneto; electrical inertia effects level the curve off at high speed, but there is no voltage drop as with battery-coil ignition.*

*Fig. 7-10. Barker-Wico 6-cylinder magneto as used on the Chevrolet.*

*Fig. 7-11. (Above and Below)—The Scintilla-Vertex 6-cylinder magneto.*

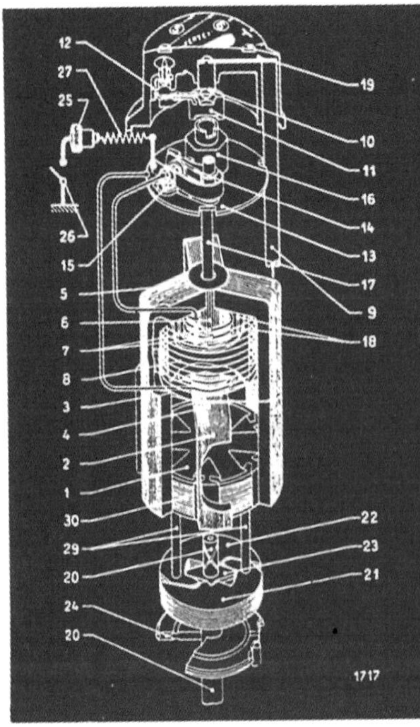


(1) Rotating permanent magnet.
(2) Short pole shoe.
(3) Long pole shoe.
(4) Core connecting piece.
(5) Core connecting piece.
(6) Armature core.
(7) Primary winding.
(8) Secondary winding.
(9) High tension distributor lead.
(10) Central contact in distributor head.
(11) Distributor rotor.
(12) Electrodes in distributor head.
(13) Contact breaker base plate.
(14) Contact breaker rocker arm.
(15) Fixed contact point carrier.
(16) Cam.
(17) Rotor spindle.
(18) Condenser.
(19) Distributor head.
(20) Driving spindle with brake.
(21) } Centrifugal weights for the
(22) } automatic advance mechanism.
(23) Cam for the automatic advance mechanism.
(24) Friction brake.
(25) Short circuiting terminal "P".
(26) Ignition switch.
(27) Resistance.
(29) Guide pins for centrifugal weights.
(30) Laminated rotor.

installation, the magneto appears to be a perfect answer to our souping ignition problem. But since the dual breaker-coil system can more than handle the job, is relatively trouble-free, and since a mag costs two or three times as much, we could only recommend one for the most severe competition conditions.

Two companies at present, Vertex and Barker-Wico, supply complete magneto units ready to bolt on the Chevrolet block (see Figs. 7-10 and 11). These are stock commercial units with special adapter equipment, which greatly simplifies the parts problem. Prices range roughly from $60 to $110.

## SPARK PLUGS

Any spark plug that can be stuck in the head can handle our firing—for perhaps three minutes! But if we want a plug that will stay right in there blasting for a 20-lap feature or for 20,000 miles on the road, we'll have to be as careful about selecting it as we are about heads, cams, and distributors.

Plugs are a critical factor in engine operation. In souping work, the problem becomes more acute because of the very high temperatures developed in the cylinder. With a full house and 10:1 compression, peak gas temperature during combustion at full throttle might reach some 4,500° F. Under a blast like this, a stock plug will quickly heat up to a point where it will touch off the fuel mixture on the compression stroke long before the spark is timed to fire. This is called "pre-ignition", and it's deadly to engine parts as well as HP. Not only this, but an overheated plug will burn out in a hurry.

So we see that the ability of a plug to absorb and get rid of heat is a vital factor—and we can't necessarily depend on the stock type in a souped engine. The "heat range" of a plug can be easily controlled by varying the length of the heat flow path from the tip around through the body and gasket to the cylinder head. In other words, by shortening this distance through which the heat must pass, we decrease the flow resistance, increase the heat flow rate, and allow the plug tip to run cooler at a given cylinder temperature. (But also remember that we must not have plugs running too cool, or they will become fouled with oil at light loads).

Fig. 7-12 shows one type of Champion plug in seven different heat ranges. For the "cold" or "hard" plugs on the left the flow path is very short, graduating to the hotter or "soft" plugs on the right. Selecting the proper plug for a souped engine is simple with such a wide range to choose from.

Before 1941, Chevrolet used a 14-mm. plug; in that year, with the introduction of the new head, they went to 10-mm. types. In 1949 they changed back to 14-mm.

The latter size is very well suited to high-output conditions, but even the coolest 10-mm. jobs will literally burn up under extended full-throttle running at high compressions. These should be replaced on all but the mildest souping jobs. The best way to do this is merely to drill out the spark plug holes with a ½-in. drill (12.7 mm.) and tap to the 14-mm. thread. An easier way is to use the special adapters now on the market; these are simply sleeves tapped to take the 14-mm. plug at one end and to

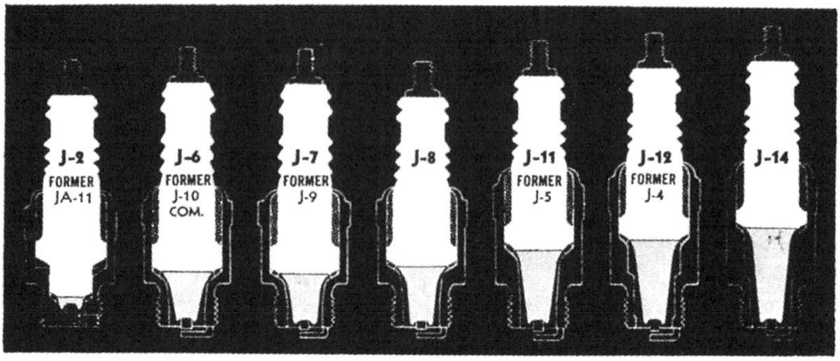


*Fig. 7-12. Drawing illustrating spark plug "heat range" (on the Champion 14-mm type); cold on the left, graduating to hot on the right.*

PROPER HEAT RANGE APPLICATION after appreciable operation, is shown below:

REGULAR FUELS

Rusty brown to grayish tan powdery deposit on firing end of plug; normal degree of electrode erosion.

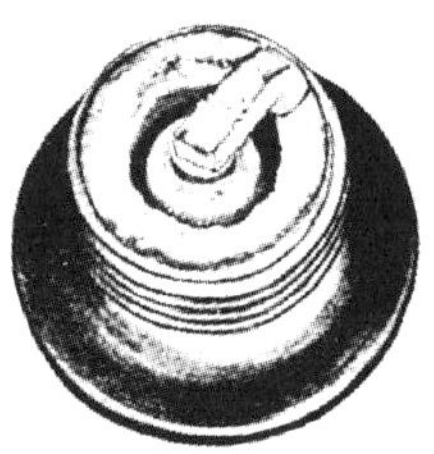

LEADED FUELS

White powdery or yellowish glazed deposit on firing end of plug; normal degree of electrode erosion.

NOTE: Deposits characteristic of leaded fuels, sometimes called "encrustments", do not interfere materially with spark plug operation, and should merely be cleaned off at regular service intervals.

WORN-OUT PLUG

Caused by normal service beyond life of plug. Spark plugs should be replaced every 10,000 miles for maximum economy and performance.

PLUGS TOO COLD for the type of service are shown below:

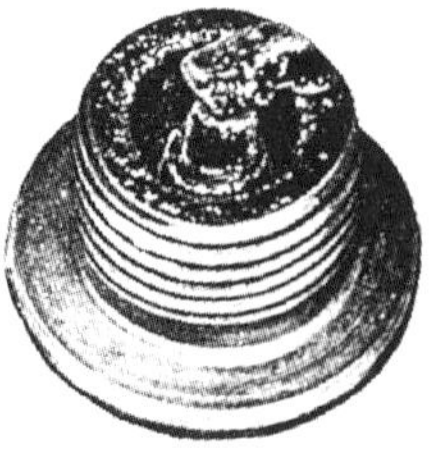

OIL FOULED

Wet, sludgy deposit on firing end of plug; negligible degree of electrode erosion.

GAS FOULED

Dry, fluffy deposit on firing end of plug; minor degree of electrode erosion.

PLUGS TOO HOT FOR THE TYPE OF SERVICE GENERALLY APPEAR AS FOLLOWS:

White, burned or blistered insulator nose; badly burned and corroded electrodes.

*Fig. 7-13. Photos showing the usual appearance of the plug tip under different engine operating conditions.*

thread into the 10-mm. hole at the other. These restrict heat flow from the plug and give worse results than straight 10-mm. plugs. We don't recommend them.

As for plug heat range, the late stock Chevy uses an AC "46" model. This is fine for mild souping; for hot road engines, the usual practice is to start colder with the AC "44" or "45" and experiment downward (or if in the Champion brand, move toward the J-6 and J-3). Consult the accompanying plug tables for comparisons of the different brands and heat ranges.

To check plug selection, about all you can do is to run a road test similar to the test outlined in Chapter 6 for checking carb jets. Accelerate hard up to high speed, declutch, and shut the motor off; if the plug base is sooty and wet, you need a hotter plug, and if glazed and burned, a colder type. Fig. 7-13 will help you here.

A word about the important item of spark gaps. This is critical because the voltage at which the spark "lets go" or jumps increases both with gap

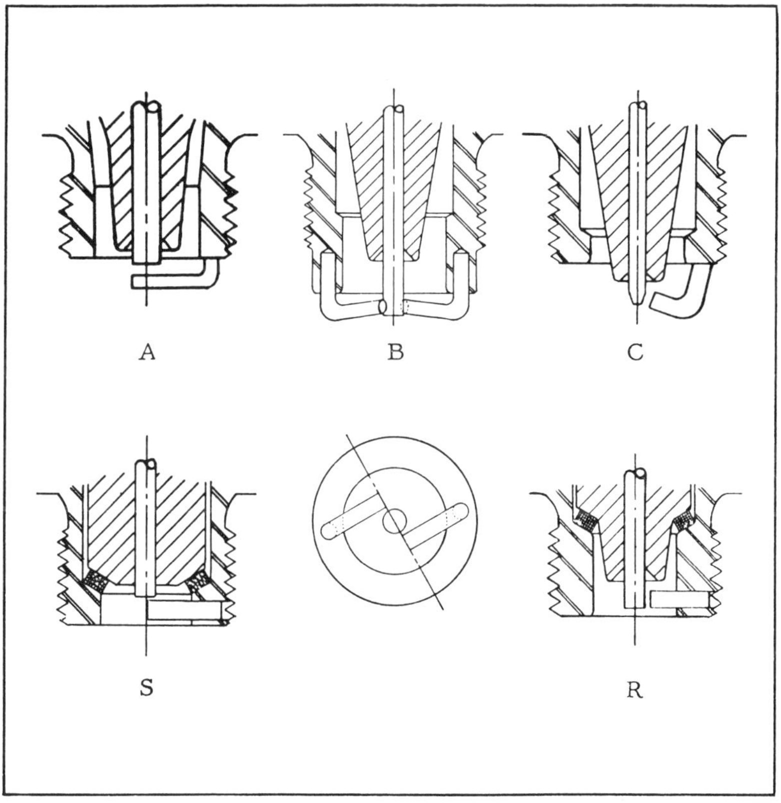


*Fig. 7-14. Champion plug electrode types for automotive engines.*

# Champion Automotive Plug Tables

| Thread Size | Heat Range | Plug Type | Hex. Size | Reach | Constr. | Stand. Gap | Electr. Type | Chief Applications |
|---|---|---|---|---|---|---|---|---|
| 18mm | Hot | R-3 | 1 | ½ | 2-piece | .015-.018 | R | Offenhauser, Miller, Riley, Bugatti engines. |
| | ↑ | R-7 | 1 | ½ | 2-piece | .019-.022 | R | Mercedes. Offenhauser, Miller, Riley engines. |
| | | R-1 | 1 | ½ | 2-piece | .015-.018 | R | Riley, Alfa-Romeo engines; Gray, Lycoming, Elto engines. |
| | | R-11 | 1 | ½ | 2-piece | .015-.018 | R | Offenhauser, Miller engines; Gray, Lycoming inboard engines. |
| | | R-11S | 1 | ½ | 2-piece | .011-.014 | R | Outboard racing. |
| | ↓ | R-2 | 1 | ½ | 2-piece | .015-.018 | R | Miscellaneous racing engines. |
| | Cold | R-2S | 1 | ½ | 2-piece | .011-.014 | R | Outboard racing. |
| 18mm | Hot | R-15 | 1 | ¾ | 2-piece | .015-.018 | R | Alfa-Romeo and miscellaneous engines. |
| | ↑ | R-16 | 1 | ¾ | 2-piece | .015-.018 | R | Alfa-Romeo and miscellaneous engines. |
| | | R-17 | 1 | ¾ | 2-piece | .015-.018 | R | Novi, Thorne engines. |
| | ↓ | R-18 | 1 | ¾ | 2-piece | .015-.018 | R | Novi engines. |
| | Cold | RJ-19S | 1 | ¾ | 1-piece | .015-.018 | S | Masserati, Alfa-Romeo engines. |

*Fig. 7-15. Champion spark plug tables.*

| Thread Size | Heat Range | Plug Type | Hex. Size | Reach | Constr. | Stand. Gap | Electr. Type | Chief Applications |
|---|---|---|---|---|---|---|---|---|
| 7/8″-18 | Hot | 3 Com. | 15/16 | 3/4 | 1-piece | .025 | A | Kerosene, Distillate tractors, oil field engs., older trucks. |
| | ↑ | 2 Com.L | 15/16 | 3/4 | 2-piece | .030 | B | (same as 3 Com.) |
| | | 22 | 15/16 | 5/8 | 2-piece | .025 | A | Buda engines; miscellaneous gas engines. |
| | | C-4 | 1 1/8 | 5/8 | 2-piece | .025 | C | Ford equip. mdls. A & B; other cars thru 1933. |
| | | 6 | 7/8 | 5/8 | 1-piece | .025 | C | Willys-Knight; Buicks thru '28; old mdls. Reo, Nash. |
| | ↓ | 1 Com. | 15/16 | 5/8 | 1-piece | .025 | A | John Deere & miscell. tractors; Autocar & miscell. trucks. |
| | Cold | 0 Com. | 15/16 | 5/8 | 1-piece | .025 | A | I.H.C. equip.; miscellaneous trucks and tractors. |
| 18mm | Hot | 9 Com. | 1 | 5/8 | 1-piece | .025 | A | Miscellaneous tractors and industrial machinery. |
| | ↑ | C-15 | 7/8 | 1/2 | 1-piece | .025 | C | Buicks 1929 thru 1937; for tractors fouling 15-A. |
| | | C-7 | 1 | 1/2 | 2-piece | .025 | C | Miscellaneous passenger cars thru 1933. |
| | | 8 Com. | 1 | 1/2 | 1-piece | .025 | A | Autocar, Brockway, Diamond T trucks; B & S and miscell. eng. |
| | | 15-A | 7/8 | 1/2 | 1-piece | .025 | A | I.H.C. equip., LeRoi and miscellaneous engines. |
| | | 7 | 1 | 1/2 | 1-piece | .025 | A | Ford V-8 equip. 1934-'37; G.M. cars 1932-'36; miscell. outboards. |
| | | 13 | 1 | 1/2 | 1-piece | .020 | R | Hall Scott & miscell. engines; miscell. passenger cars. |
| | | 6 Com. | 1 | 1/2 | 1-piece | .025 | A | Brockway, Corbitt, Reo, White, Federal & miscellaneous trucks. |
| | | H-17-A | 7/8 | 1/2 | 1-piece | .025 | A | I.H.C. power units & farm machinery; miscellaneous engines. |
| | ↓ | 5 Com. | 1 | 1/2 | 1-piece | .025 | A | I.H.C. trucks; miscell. trucks severe service conditions. |
| | Cold | H-16-A | 7/8 | 1/2 | 1-piece | .025 | A | Hall Scott engines and miscellaneous applications. |
| 14mm | Hot | J-14 | 13/16 | 3/8 | 1-piece | .037 | A | For Buicks and others fouling J-12. |
| | ↑ | J-12 | 13/16 | 3/8 | 1-piece | .037 | A | Buick after 1937. |
| | | J-11 | 13/16 | 3/8 | 1-piece | .037 | A | Chrysler Corp. cars 1932-'36. |
| | | J-8 | 13/16 | 3/8 | 1-piece | .025-.028 | A | Chrysler Corp. cars after '36; Chevrolet 1937-'40. |
| | | H-10 | 13/16 | 7/16 | 1-piece | .025-.028 | A | Ford, Lincoln, Mercury after '37; most H. C. aluminum heads. |
| | | J-7 | 13/16 | 3/8 | 1-piece | .025-.028 | A | Kaisers, Frazers; Nash after 1942; Willys, Studebaker. |
| | | H-9 Com. | 13/16 | 7/16 | 1-piece | .025-.028 | A | Ford Commercial Cars after 1937; Miscellaneous engines. |
| | | J-6 | 13/16 | 3/8 | 1-piece | .025-.028 | A | Graham 1940-'41; miscellaneous commercial cars and buses. |
| | ↓ | J-3 | 13/16 | 3/8 | 1-piece | .025-.028 | R | Outboard engines; miscellaneous racing engines. |
| | Cold | J-2 | 13/16 | 3/8 | 1-piece | .015-.018 | R | Miscell. trucks; Mack fire equip.; Gray marine & racing. |
| 10mm | Hot | Y-8 | 5/8 | 1/4 | 1-piece | .037-.040 | A | For cars fouling Y-6; Chevrolet cars after 1940. |
| | ↕ | Y-6 | 5/8 | 1/4 | 1-piece | .037-.040 | A | Chevrolet trucks after 1940; Packards 1937-'42. |
| | Cold | Y-4-A | 5/8 | 1/4 | 1-piece | .028-.030 | A | Packard equipment after 1942; Cadillacs, LaSalles. |

# Spark Plug Range Comparisons

| | | A<br>Extremely Hot | B<br>Very Hot | C<br>Hot | D<br>Warm | E<br>Cool | F<br>Cold | G<br>Very Cold | H<br>Extremely Cold | | |
|---|---|---|---|---|---|---|---|---|---|---|---|
| AC | 10 mm | 108<br>M8 | | 106 | | 104 Com<br>104 | | 103 Com | | 10 mm | AC |
| | 14 mm | 48 | 47<br>47 Com | 46-5 46<br>46 Com | 45 45L Com*<br>45 Com 45L* | 44 Com<br>44-5 Com 44 | 43L*<br>43L Com* | 43 Com 42-5 Com | 42 Com | 14 mm | |
| | 18 mm | 88S Com<br>88L Com* | 88 87S Com<br>87 Com | 86<br>86 Com | 85 Com<br>85S Com | 84 83S Com<br>83 Com | 82 Com<br>82S Com | | 81S Com | 18 mm | |
| | 7/8" | 78L Com* | 77 Com 77L Com*<br>78 78S | 76 76 Com<br>76S | 75 Com | | 74 74 Com | 73 Com | | 7/8" | |
| CHAMPION | 10 mm | Y-8 | | Y-6 | | Y-4A | | | | 10 mm | CHAMPION |
| | 14 mm | J-14 | J-12 | J-11<br>H-11* | J-8<br>H-10* | J-7 | J-6<br>H-9 Com* H-8* | | J-2 | 14 mm | |
| | 18 mm | 9 Com* | C-7<br>C-15 | 8 Com | 7<br>15A | 13<br>6 Com | 5 Com | R-7 R-1 | R-11 R-2 | 18 mm | |
| | 7/8" | 3 Com*<br>22 | 2 Com L* | C-4 6 | | 1 Com | 0 Com | | | 7/8" | |
| AUTO-LITE | 10 mm | P6<br>PR6 | | | | P4<br>PR4 | | | | 10 mm | AUTO-LITE |
| | 14 mm | A11<br>AR10 | | A9 AR8<br>ARL8* | A7 AN7<br>AL7* | | | A5 AR5 AN5<br>AL5* ARL5* | A3 | 14 mm | |
| | 18 mm | B11 | | B9 BR8 | B7 | B7 | B5<br>BR4 | | | 18 mm | |
| | 7/8" | T11 | | T9 | T7 | T7 | | | | 7/8" | |
| EDISON | 10 mm | 2-S | | 3-S | | 4-S | | | | 10 mm | EDISON |
| | 11 mm | | 52-S | 53-S | 55-S L-55-S | 56-S | L-56-S | 57-TS 58-S | 59-TS 58-S | 14 mm | |
| | 18 mm | 42-T | Z-19 43-S<br>Z-146-S | 43-TS | 44-HS<br>Z-147-S | 45-S<br>46-TS | 48-TS | 49-S | 49-TS<br>Z-149-S | 18 mm | |
| | 7/8" | 31-T Z-162-S | X-46 | 35-S | 37-TS | 38-TS | 39-TS | 40-TS | | 7/8" | |
| GLOBE UNION INC. | 10 mm | G0-110 | | G0-165 | | | G0-230 | | | 10 mm | GLOBE UNION INC. |
| | 14 mm | G4-85 | G4-120 | G4-150 | | G4-185 | G4-220 | | | 14 mm | |
| | 18 mm | G8-85 | G8-125 | G8-165 | | G8-195 | G8-220 | | | 18 mm | |
| | 7/8" | G7-110 | | G7-150 | | G7-175 | G7-195 | | | 7/8" | |
| FIRESTONE | 10 mm | T-60-F | | T-40-F | | T-40-F | | T-20-F | | 10 mm | FIRESTONE |
| | 14 mm | F-120-F | | F-80-F | F-90-LF* | F-40-F | F-50-LF* | F-30-F | | 14 mm | |
| | 18 mm | M-120-CF | | M-80-CF | | M-40-CF | | | | 18 mm | |
| | 7/8" | S-120-CF | | S-80-CF | | S-40-CF | | | | 7/8" | |

TORQUE WRENCH CHART

Always use a spark plug socket wrench or a torque wrench. These wrenches are readily obtainable and are the only kind which will avoid distortion of the plug and insure the insulator against damage or breakage.

*Long reach.

Average torque wrench pressures recommended for standard plugs in vehicles. All pressures listed are based on spark plug and engine threads being clean.

| Plug Thread | Cast Iron Heads | Aluminum Heads |
|---|---|---|
| 10 mm | 14 lb-ft | 12 lb-ft |
| 14 mm | 30 lb-ft | 28 lb-ft |
| 18 mm | 34 lb-ft | 32 lb-ft |
| 7/8" | 37 lb-ft | 35 lb-ft |

*Fig. 7-16. Heat range comparisons for various plug brands.*

and compression pressure. Using large stock gaps (0.035 in.) with 8:1 or more compression increases the punch at the electrodes to the point where the plug will fail in a few-thousand miles. A smaller gap is the answer here — but there is a catch: At low RPM when gas turbulence in the cylinder is at a minimum, combustion lags with very short gaps, which of course, hurts low-speed torque output and even causes irregular firing at idling. So we must consider low-speed performance in spark setting.

Here are the recommendations: For very mild souping jobs, just use the stock gap of 0.035 in. For heavier souping, but for general city and highway driving, set the gaps back to 0.020-0.025 in. As for competition, we can't worry too much about long-wearing; for "drag" racing from a slow start, 0.025-0.030 gaps will give best acceleration, but the plugs won't take it for long. For track racing where you don't need to worry about low-speed output, your plugs will stay longer and be more dependable if you use gaps of 0.015-0.020 in.

## TIMING

Spark advance is of prime importance in getting that last HP, as we learned in Chapter 3. Since the fuel-air mixture doesn't burn instantly, we must advance or retard the spark under different conditions of load and speed to obtain the maximum effective cylinder pressure on the power stroke.

Actually, spark timing will be no great problem on the Chevy because the stock advance "curve" has been found to work out very well under souped conditions. The centrifugal linkage here gives an advance increasing from about 3° at idling speed to 38° at 3500 rpm; the vacuum linkage will give up to 20° additional advance at the lower speeds. For road engines, this will do very well. (For track racing at speeds in the vicinity of 5000 rpm with the low-turbulence Wayne head, you may need 50-60° advance; the special or dual magneto ignitions which you would be using in this case can be adjusted to this range.)

Incidentally, here's a tip: If you have reworked your cylinder head for high compression and are bothered with severe knock at low speeds, you can often fix things up with a flick of the "Octane Selector" bolt on the distributor (see Fig. 7-17). This allows up to 10° advance or retard over the stock timing curve by loosening the bolt and rotating the dial to the

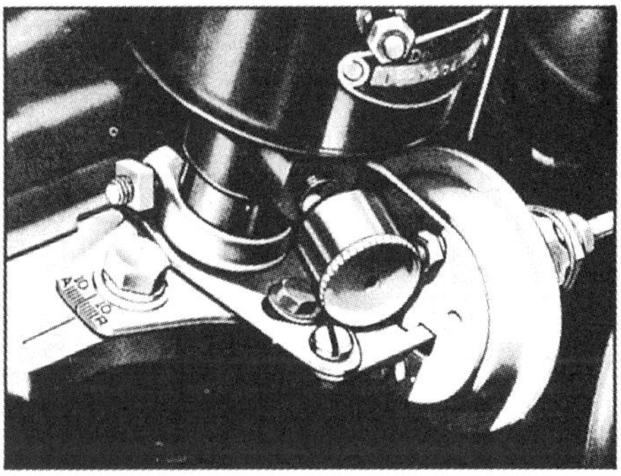

***Fig. 7-17. Octane selector adjustment on the Chevy distributor, allowing 10° spark advance or retard over the stock advance curve.***

desired position. Retard the spark until you only get a trace knock—but remember that every degree of retard kills off HP at the top end, so don't overdo it.

A better answer to this low-speed knock on acceleration, if you don't mind a little experimenting, is to grind down the centrifugal flyweights in the distributor. This won't decrease your overall advance at high RPM, but only the rate of advance during acceleration. Make sure you grind both arms to the exact same contour and weight.

We suggest you think this problem over carefully in your souping plans. Proper ignition is not an overly expensive factor, but it's absolutely necessary to the success of any souped engine. Don't forget it!

Now let's skip off the beaten souping path for a bit and have a look at another weird and wonderful way of boosting HP that has hardly been touched by hot-rodders: Supercharging. As we learned in Chapter 3, the power output depends on the weight of fuel-air mixture we burn per minute. Getting it with carbs and cams is alright—but a little blower pressure makes Mother Nature's atmospheric pressure look pretty sick when it comes to packing it in in king-size chunks!

## SUPERCHARGERS

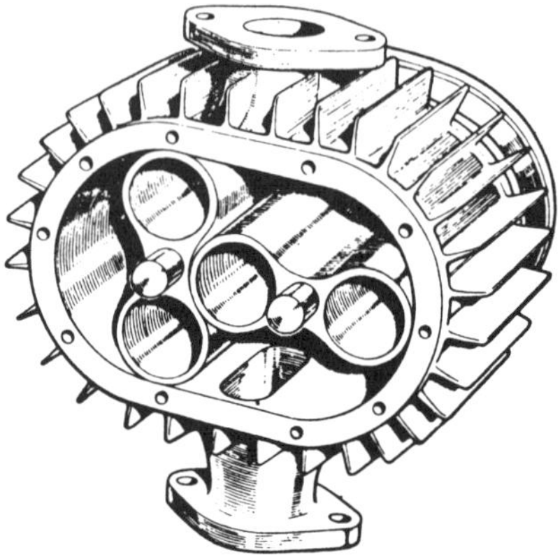

The Roots-type supercharger. There is a fine clearance between the two geared paddles and between these rotating units and the casing.

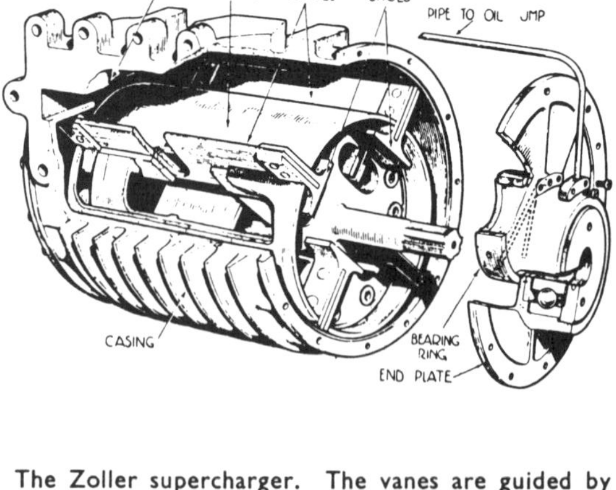


The Zoller supercharger. The vanes are guided by shoes and bearing rings so that they do not rub against the casing. Note the cooling fins on the casings which are necessary to carry away the heat generated in action.

# CHAPTER 8
## SUPERCHARGERS

WHAT U. S. stock car engine can claim the distinction of being the first to have a supercharger available for it as specialty equipment?

Surprise! Not the Ford V8 or Chevrolet. In 1926, the Green Engineering Works of Dayton offered a centrifugal blower for the Model "T" Ford, geared up 7:1 and pumping 6-7 lbs./sq.in. boost; it jumped the output of the old "Lizzie" from a modest 20 hp at 1600 rpm to a staggering 30 hp at 2000 rpm!

Yes, even back in the days of the Frontenac head and the Miller carburetor, the boys in the "know" realized the terrific potential of pressure induction as applied to the stock engine. Here, at last, was a way to decrease the burdensome lbs. per HP of those bulky power plants on pump gas, without the attendant fuss and bother of noisy valves, constant knock, open exhaust, poor idling, etc.

Since then, we've come a long way. Auburn, Dusenberg, Cord and Graham developed practical units for their stock models—and in 1937, McCulloch brought out a specialty blower for the Ford V8. The Chevrolet was forgotten until a couple of years ago, but we now have a couple of units available—and there will be more later.

Make no mistake, supercharging is a beautiful and practical souping weapon. Theoretically there's no limit to the amount of HP you can pull by just putting more and more pressure to the intake manifold. That stands to reason from what we've learned about performance and the weight of charge inducted on the intake stroke. In practice, of course, things don't run along quite according to theory and there are a thousand items that hold us back. But the fact remains that we're working with something (pressure) that has no top limit—ingenuity, engineering, and money can give us about any HP we want from it. We'd like to see a lot more work with supercharging, rather than all this fiddling with "super fuels". It would pay off more in the long run, in performance as well as sound engineering progress.

### SUPERCHARGING PRINCIPLES

In Chapter 4, we stated that a basic aim in souping is to increase the weight of fuel-air mixture inducted into the cylinder on the intake stroke. This enables us to burn a greater weight of fuel in a given time and increases HP accordingly. Obviously if we can maintain the mixture at a positive *pressure* in the manifold, we get a proportionately greater weight of it into the cylinder than would be sucked in under atmospheric pressure (14½ lbs./sq.in.) —and we get more HP at all RPM. This is what a supercharger does. It's nothing more than a large-capacity fluid pump designed to maintain a positive pressure in the intake manifold at all speeds (with full throttle, of course).

But here's the catch: There are two distinct types of superchargers available for the Chevrolet engine today—and thereon hangs a vital tale. It's all in the way the mixture is compressed. There are (1) the "centrifugal" type, such as the Besasie, which compresses by virtue of fluid momentum, and (2) the "positive-displacement" or Roots type, such as the Italmeccanica, which simply pumps the fluid.

## THE CENTRIFUGAL TYPE

Fig. 8-1 shows the general layout of this type. It consists merely of a rotor or "impeller" disk with a number of radial blades on one side, rotating in a casing. The fuel-air mixture is scooped in at the hub and whirled by the radial blades; centrifugal force pulls it toward the outside edge of the disk, where it is hurled off into the outlet casing at high velocity. From here it literally piles up in the intake manifold – in other words, the momentum energy of the high-velocity gas flow is converted into pressure energy when the gas loses speed in the manifold. In the case of the automobile supercharger, we turn the impeller 20,000-30,000 rpm at peak speed, which gives an exit gas velocity off the impeller of something around 800 ft./second, and a final pressure of 3-6 lbs., depending on blower design.

We know that the momentum energy of any given mass increases as the square of its velocity. That's a basic physical principle. This is just what happens with our centrifugal, or "fluid momentum" supercharger – the full-throttle pressure output falls off as the square of the impeller RPM. With the impeller geared or belted to the crankshaft, this makes a very bad situation for a road engine; by driving the impeller with an exhaust turbine, things are much better.

*Fig. 8-1. Diagram of a centrifugal supercharger.*

## THE ROOTS TYPE

This type is named for its inventor and the general layout is shown in Fig. 8-2. It compresses the fluid in an entirely different way than the centrifugal type. In this case, the gas is caught up on the inlet side by theinterlocking rotors, carried around between rotors and casing, and simply discharged into the manifold. There is no compression inside the blower at all — it just pumps gas faster than the engine can burn it, and maintains a pressure by virtue of piling up fluid in the manifold until a state of equilibrium is reached.

*Fig. 8-2. Cross-sectional drawing of a typical two-lobe Roots-type supercharger; arrows show the air flow path.*

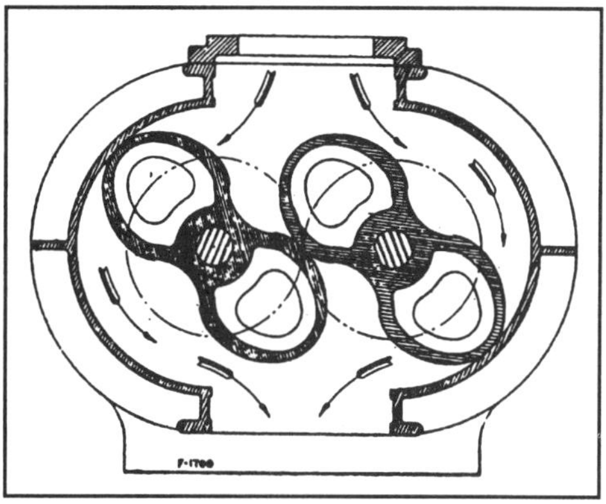


From this it would seem that pressure output here would be constant at all speeds, but this is not quite true. In order to eliminate noise, friction, and wear in the Roots-type blower, it is constructed to run with a small clearance of about 0.005 in. at all points. Thus there is always a "slip loss' or leakage of gas back past the rotors, the amount depending only on the boost pressure ,and not RPM). As a result, since this leakage forms a larger proportion of the total discharge at low speeds, we find that the "volumetric efficiency" of a Roots blower falls off from around 85% at peak speed (4000-8000 rpm in usual sizes) to maybe 65% in the neighborhood of 1000 rpm of the engine. This is not too bad — and is certainly far superior to the centrifugal type from this standpoint.

Fig. 8-3 shows approximately the full-throttle manifold pressure curves we'd get with the two types of superchargers, with both turbo and gear drive on the centrifugal, if all layouts are designed to give a boost of 6 lbs. at 4500 rpm. This shows clearly the superiority of the positive-displacement type for low-RPM output.

## CHOOSING A TYPE

Besides this all-important factor of pressure output, there are several other performance items that should influence our decision on whether to buy a centrifugal or Roots type. Here is a listing of the important ones:

*Mixture heating* — Compression generates heat. Compressing the fuel-air mixture in a supercharger raises its temperature, and this aggravates the engine-cooling problem and increases the fuel octane requirements. Since the centrifugal type compresses within the casing, the heat developed goes

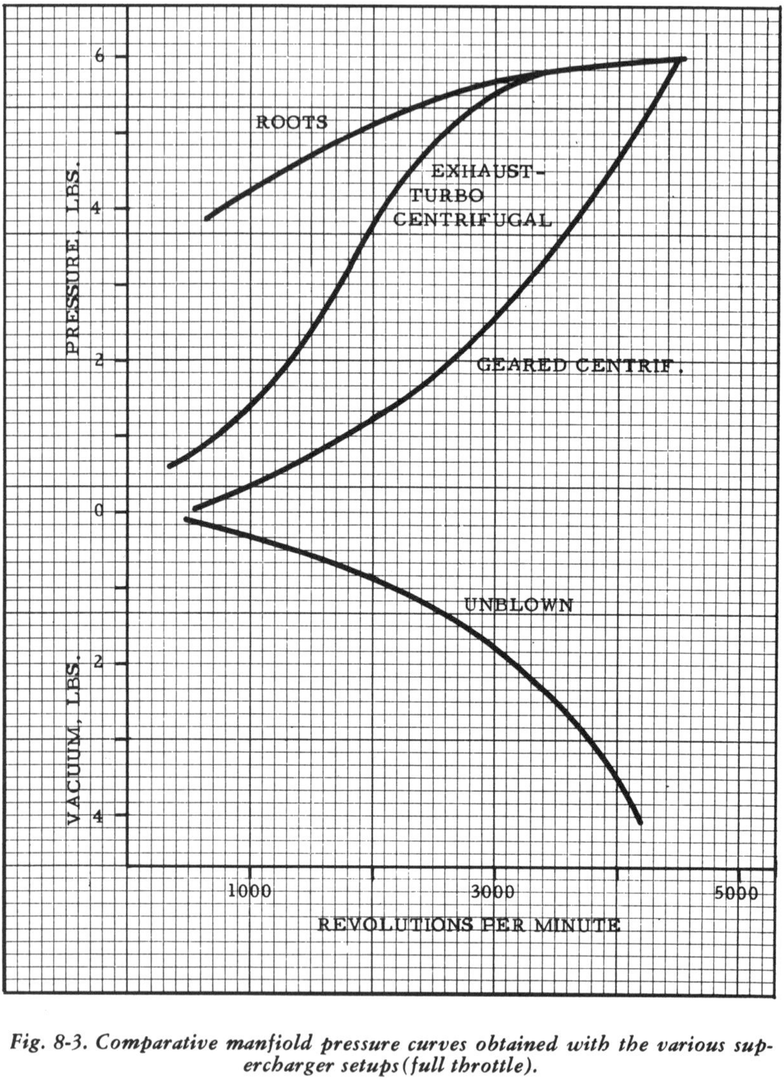


*Fig. 8-3. Comparative manfiold pressure curves obtained with the various supercharger setups (full throttle).*

into additional pressure energy (just like the compression in the cylinder — remember?). The Roots, on the other hand, does not compress internally at all — so none of its compression heat becomes pressure energy.

For example, if both types compressed pure air from room temperature (70° F.) to say 5 lbs./sq.in., final temperature would be about 140° with the centrifugal and 250° with the Roots. As a result, things like engine cooling, fuel octane, and spark timing will be somewhat more critical with the Roots type.

*Blower power loss* — For the same reasons outlined above — that is, compression heat rise — the Roots type will also require more power to pump a given flow rate because no compression heat is going into useful pressure energy. This will cut our available HP at the flywheel slightly on a given fuel octane limitation, and decrease gas mileage.

*Fuel consumption*—Any supercharger imposes a whipping motion on the ingoing fuel-air mixture, which gives much more even distribution between cylinders. Thus all cylinders can run at a fairly efficient mixture ratio, instead of the wide 50% variations we find in unblown engines — and we therefore burn less fuel for a given HP. The centrifugal type is better in this respect than the Roots, due to the high impeller tip speed.

*Reliability* — Here the Roots has it over the centrifugal. Since the centrifugal runs at about 6:1 with the crank, violent engine acceleration or deceleration such as we get when we "goose around," accelerates the blower at six times that rate and imposes terrific stresses on the gears and bearings. This tends to cause faster wear and increases the possibility of mechanical trouble.

The Roots, on the other hand, runs at much lower speed (nearer crank speed) and can be built very rugged all the way through. Actually, however, modern supercharger design is so well advanced that breakdown is not much of a problem on either type any more.

With this information, the job of selecting a type for our various souping purposes is simple: If you are designing a fast road car for long-distance highway travel, with good mileage, very high cruising speed, and plenty of "dig" between 50 and 90 mph, the centrifugal is your answer. If you're planning a street job that won't be used primarily for long high-speed runs, but will have back-breaking acceleration at *all* speeds, the centrifugal, even turbo drive, is not too good — and the Roots is your answer. For any type of road racing, the Roots should always be used. For high-RPM track racing in general, the centrifugal is better, but due to the excessive dirt in the air on most tracks, supercharging is not too popular here.

Let's take a quick look at the supercharging equipment that is presently available for the Chevy block:

## BESASIE

Bob Besasie, owner of the Besasie Engineering Co., of Milwaukee, is one boy who can't stand the idea of tying the centrifugal supercharger to the crankshaft by a belt or gears. He wants that impeller to be able to run relatively faster at low engine RPM to make up for some of that killing pressure loss suffered by this type of blower at low speeds. An exhaust-turbine drive is the obvious answer.

The theory is this: Since the "breathing", or weight of fuel-air charge drawn in on the intake stroke, is at a maximum at moderately low speeds, this means that the *weight* of exhaust gas flow *per revolution* will vary in proportion. Obviously this gas flow can be utilized to drive a blower impeller through a turbine wheel — at a speed ratio varying with throttle opening and crank RPM!

Besasie has been experimenting with this idea (applied to automotive engines) as a hobby for some fifteen years, and has recently come up with a unit which he believes to be the answer. Although the resulting blower boost curve is inferior to the positive-displacement type over the full speed range, it is greatly superior to the gear or belt-driven centrifugal — and gets rid of most of the drive friction to boot. The whole thing looks beautiful from a theoretical standpoint.

The critical problem, of course, is exhaust back-pressure. At peak speed, the turbine sets up a resistance of 3½ lbs./sq.in., or about what the stock exhaust system gives. Downstream of the blower, it is vital that not over ¼ lb. additional resistance be present — or the exhaust valves are goners. Besasie specifies 2¼-in. tubing and a straight-through muffler at least 2 in. inside diameter (although due to the gas energy loss in the turbine, you don't really need a muffler).

One purchaser disregarded the recommendations and fitted a 1½-in. line and stock muffler; the car would go to 50 mph from a dead stop in 6½ seconds, but the acceleration flattened right out at 75. Besasie checked the

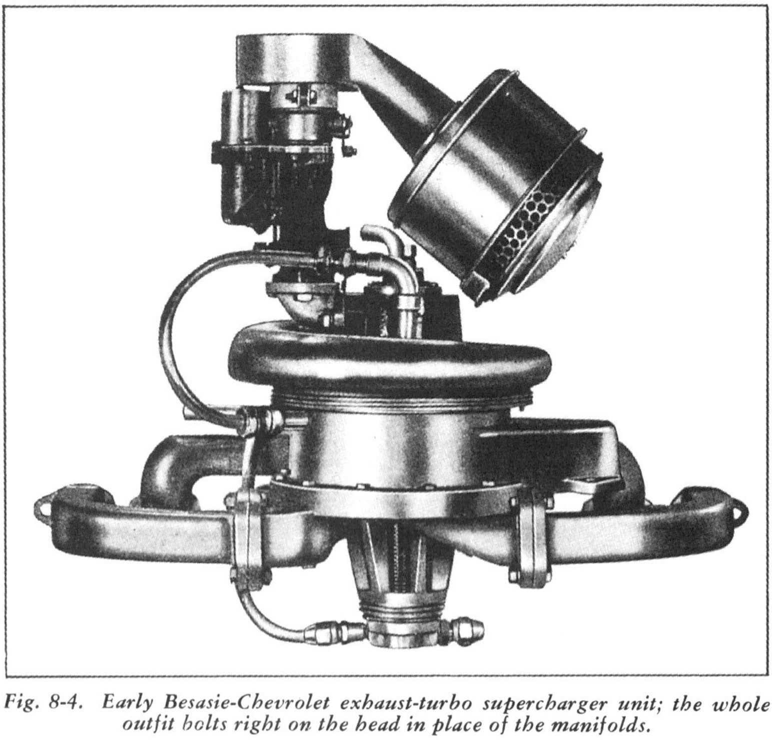

*Fig. 8-4. Early Besasie-Chevrolet exhaust-turbo supercharger unit; the whole outfit bolts right on the head in place of the manifolds.*

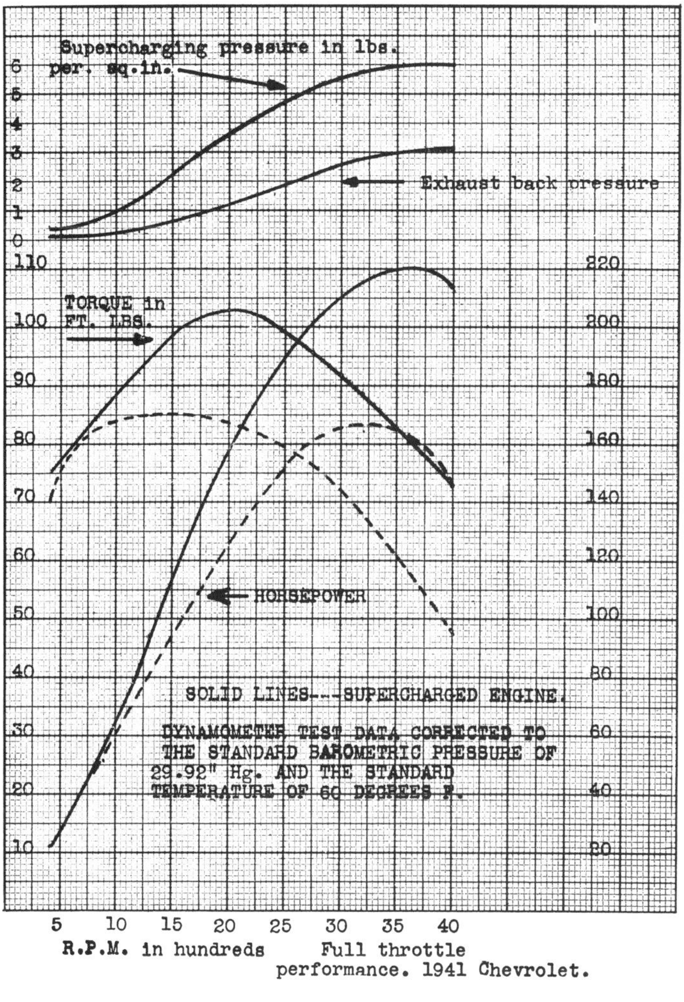


*Fig. 8-5. Performance curves for the early Besasie blower on a '41 Chevy engine.*

installation later and measured the back-pressure at 6½ lbs.! So remember this matter in relation to the exhaust-turbo supercharger. On the other hand, if the thing is set up right, the road performance is sensational and it will definitely stand up. Besasie has run one 157,000 miles in his '40 Chevy and can out run anything stock on the road with a clear 100-mph top speed.

Getting down to what there is available for you, the company brought out the first commercial model for the Chevy a couple of years ago (see Fig. 8-4). Fig. 8-5 shows performance curves for this setup on a '41 engine. The power boost of only 32% at 6 lbs. boost was not considered good, and for 1951, Besasie has introduced an improved model giving much better performance all around (changes are mostly on the intake side).

The whole unit bolts right on the head in place of the intake and exhaust manifolds. Fig. 8-6 shows the latest installation and Fig. 8-7 gives dynamometer curves as mounted on a '51 Powerglide engine. That new peak of 155 hp at 4500 rpm on 6 lbs. boost needs no apology in any circle. The unit itself is very rugged and straightforward. The turbine and impeller operate on one shaft running on precision ball bearings; the impeller blades are electrically welded to the disk for maximum strength, and lubrication is provided by an oil mist fed by engine pressure.

(In addition to the special exhaust system, Besasie suggests Stellite exhaust valves, aluminum pistons, and dual ignition for best durability with this setup if you're going to "push" much. Using other methods such as boring, high compression, cams, etc., potential outputs of 250 hp or more on pump gas are entirely possible!)

The complete unit sells for around $250.

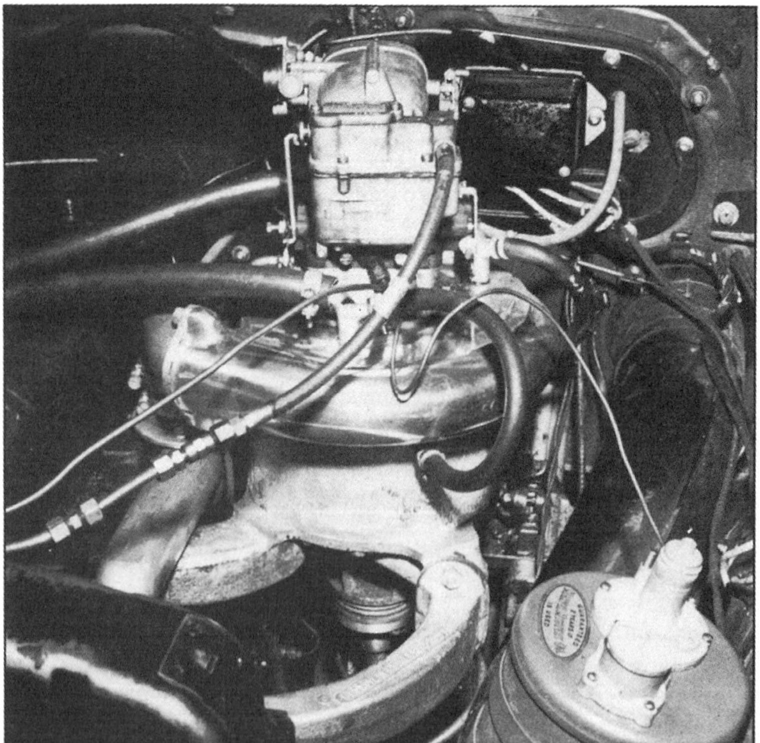

*Fig. 8-6. The latest Besasie supercharge installation on a '51 Chevy with Olds "Rocket" carburetor; note that the impeller casing is water-heated for maximum vaporizing of the ingoing charge.*

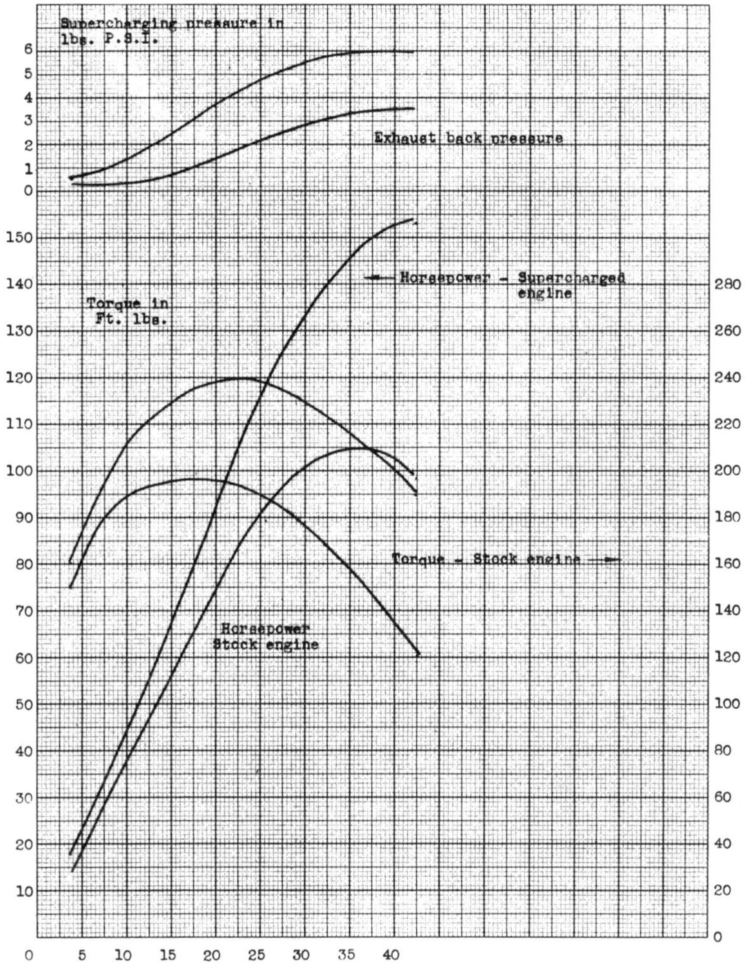


*Fig. 8-7. Performance curves for the new Besasie blower on a '51 Chevrolet Powerglide engine.*

*Fig. 8-8. A '40 Chevy coupe powered by a Besasie-converted supercharged engine in action in South American stock car road racing; the length of this race was 3,240 miles.*

*Fig. 8-9. The I. T. Roots-type supercharger installation on a Chevy engine; drive is by three V-belts, and two carbs are used to give 140 hp.*

## ITALMECCANICA

This is the well-known "I.T." blower, formerly manufactured in Turin, Italy, for a number of U.S. and foreign stock engines. The American units are now being produced at a factory in New Jersey to standard American thread sizes, etc., to simplify the maintenance problem. The Chevrolet installation is shown in Fig. 8-9, and sells for about $360 complete.

The I.T. blowers are positive-displacement pumps of the Roots type, which give a more or less constant boost pressure and provides a substantial performance increase over the full RPM range. They are much better in this respect than the fixed-drive centrifugal, and will even considerably outdo a turbo-driven centrifugal below 2500 rpm. Drawbacks are greater heating of the mixture, which aggravates knock and cooling problems, and greater blower loss (which shows up as poorer gas mileage). As fitted to the standard Chevy engine (216 cu. in.), the I.T. supercharger pumps a peak boost of 6-8 lbs. and increases peak output to about 140 hp at 4200 rpm.

From the standpoint of design, this unit is beautifully planned, with a rugged belt drive and good bearing layout. There is no lubrication connection with the engine; the rear bearings run on hub grease from grease nipples, with a small reservoir containing SAE #90 oil feeding the front bearings (this is checked every 1000 miles). The whole I.T. layout is very neat and reliable, and gives the best overall performance of the Chevy blowers.

As was mentioned, supercharging is a terrific souping tool at our fingertips — and we haven't yet scratched the surface on supercharging the stock block. While we are fooling with cams, heads, carbs and such to get 0.9 hp/cu.in., 1½ hp/cu.in. is around the corner waiting to be snatched — in the form of a 10-lb. blower boost. Sure, the costs are high — the troubles are many—but so are the HP dividends. The 300-hp souped road engine is on its way, and this is how we can get it.

So now we have finished our souping job. We have gone all the way from the oil pump to the carb after the horses. But we've only begun to use our heads. Before we stick that ball of fire into a chassis and go out and full-bore it, let's do a little figuring.

*Wayne head conversion on Chev. block in Johnson Spl. as it appeared at Indianapolis Speedway, 1951. Note fuel injection setup.*

*270 cu. in. Chevrolet-Wayne conversion powers this sprint car, the "Johnson Special". Jimmy Davies is pilot in photo.*

# CHAPTER 9

## WHAT'LL SHE DO?

A SOUPER once told us he wasn't the least interested in power curves and peak torques – he just wanted "to build 'er up and let 'er go!" So we told him what happened to Leadfoot Louie, who also isn't interested in power curves, when he dropped his first hot Chevy engine in a chassis.

It seems Louie had worked up a stock block and head into a pretty fair competition setup, and wanted to use it in a track roadster. What gear ratio should he run? This was no problem for our genius hero he just took his rear tire diameter of 30 in. and went to work with a pencil and paper – and his estimations.

He was gearing for a fast half-mile that showed roadster lap times around 23 seconds. That's 78 mph average, so he geared for this speed (forgetting that speeds go much higher in the straights). Next, he carefully estimated his peak output on 8:1 compression and pump gas as 220 hp at 6000 rpm! He had read once where the best lap time was achieved if you geared slightly *over* the peak, so he set his gearing for 6500 rpm. How about that? Well, Louie did manage to get his calculations right and came up with an overall ratio of 7.45:1. He ended up by pulling a 41-9 cog in second gear.

We won't bother to tell you what happened when he let it out on the track – but Louie is working strictly with road jobs now! So we had better get to work and pay some close attention to this long-neglected subject of true HP and RPM values.

We find that most amateur rodders are sadly lacking in fundamental knowledge of these factors. Many of them don't know any more about a torque curve than Gypsy Rose Lee, and we shudder at the thought of their being responsible for setting up a fast chassis. As a matter of fact, some close idea of the true power curve of the souped engine will be invaluable in setting up the chassis properly – for selecting gear ratios, clutch capacities, tire sizes, weight distribution, braking requirements, etc. So let's take some time out and see what we can do about estimating the performance of the souped Chevy.

In the first place, this is not going to be simple. Einstein himself couldn't slip out a slide rule, take a handful of figures like displacement, compression ratio, valve area, piston weight, carb size, timing, etc., and – flip, flip – give the exact peak HP and RPM. It can't be done. There are too many variable and unpredictable factors involved to give us any such easy way out.

We will have to rely instead on data averaged from dynamometer tests on souped Chevy engines. Fortunately, quite a few of these tests have been run and we have a fair volume of data available to correlate. Even then, it will not be simple to estimate a complete power curve, but we've come up with a method we think is as easy as possible under the circumstances; it will require locating only two points (on regular graph paper), drawing a straight line between the two points with a ruler, and then sketching in a smooth curve.

## LAYOUT FUNDAMENTALS

We learned quite a lot about the power curve back in Chapter 3 – how it was shaped and why. Now when we analyze that same HP-RPM curve mathematically (which we won't bother to do here), we find an interesting fact that will help us a lot right now, and that is: If we draw a straight line from the origin of the graph – that is, the zero point on both the HP and RPM scales – and if we draw this line *tangent* to the curve, the point where it touches the *peak torque* point. Fig. 9-1 shows what we mean.

Obviously we can just as well put this simple little fact to work for us in reverse to draw the HP curve. We merely locate the points of peak HP and peak torque, draw the straight line from the origin through the peak torque point, and then sketch in the curve – with a smooth peak at the top and tangent to the straight line. Simple? Of course, our curve around the peak and below the torque point will be pretty much a guess, but you can get a good idea of the shaping here by studying published power curves (there are plenty of them in this book).

The next job – and this is the tough one – is to estimate those peak HP and torque points.

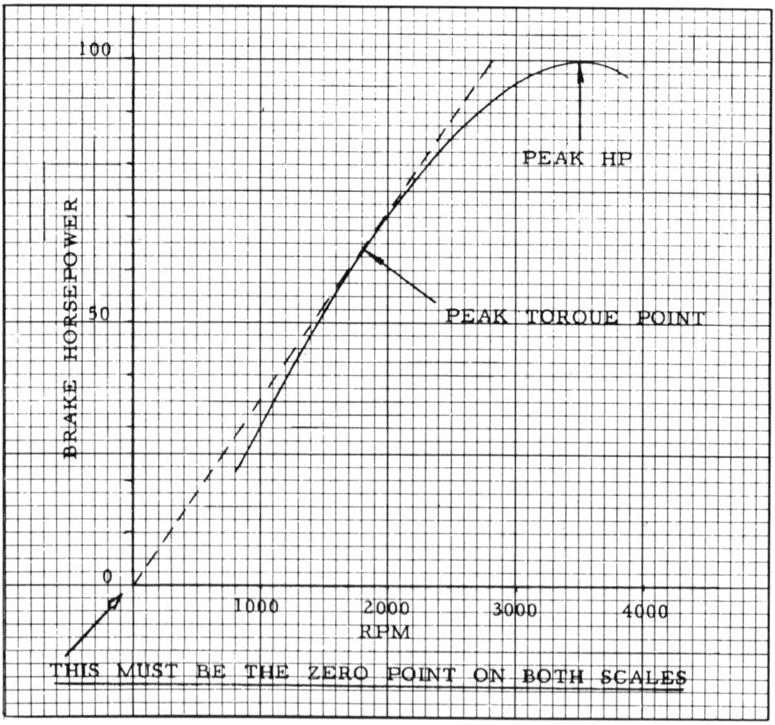


*Fig. 9-1. How to locate the peak torque point on a power curve.*

## ESTIMATING PEAK H.P.

There are several ways we might do this. In the book, *"Souping the Stock Engine"*, we used two separate methods, letting one cross-check on the other, because we were working with many widely different engine types. With just the basic Chevrolet block, however, we can simplify things a lot and get just as good accuracy, by using "HP increments" on the HP estimation, and by basing the peak RPM estimation on piston displacement and cam timing.

In the first method, we allow for the HP increase given by each individual souping step (dual manifolds, heads, cams, etc.) by assigning a definite HP increase or "increment" to each step. We then just add all the increments together and add the sum; in turn, to the stock peak HP to get the new peak. For instance, if increments for a 90-hp Chevy block are 14, 8, and 22 – then we find the total increment thus: $14 + 8 + 22 = 44$, and the estimated peak HP is $90 + 44 = 134$.

Using all available dyamometer test data, we have prepared the following table for the Chevrolet block on gasoline fuel:

TABLE I

PEAK H.P. INCREMENTS

| | |
|---|---|
| Increased compression, stock head | See Fig. 5-2, Chap. 5 |
| Wayne head | 30-40 (depending on C.R.) |
| Porting, stock head | 4 |
| Large 1⅞ intake valves, std. head | 5-6 |
| High-lift rocker arms | 2 |
| Dual manifold | 15 |
| Triple manifold | 16-17 |
| *Cams:* | |
| Semi | 13 |
| ¾ | 20 |
| Full-race | 25 |
| Super | 30 |
| Mushroom & roller | 33 |
| Increased displacement | Percent increase in displacement multiplied by 9/10 |
| Miscellaneous (ignition, oil pressure, etc.) | 4 |

For estimating peak RPM on the Chevy, the best method appears to be on a basis of piston displacement. Other factors equal, peak RPM will decrease somewhat as the piston size is enlarged by boring, or when piston speed is increased by lengthening the stroke – and since the other major factor in RPM is valve timing, we can represent peak RPM on a graph as a function of the cam grind and the displacement. We have prepared Fig. 9-2 from all available data for estimating peak RPM on a souped Chevy (this graph should not be used for other basic engines).

Let's take an example and see what we can do with the above estimation system. Suppose we have a standard 90-hp Chevy engine and do a mild road souping job; we leave compression ratio as is, but port the head and install high-lift rockers, a dual manifold, and a ¾ camshaft. What is the estimated peak?

For peak HP, refer to Table I and get the following increments: 4 for the porting, 2 for the rockers, 15 for the manifold, and 20 for the cam. Adding up, we get a final increment of $4 + 2 + 15 + 20 = 41$, and the estimated peak HP is $90 + 41 = 131$ hp. Using Fig. 9-2 for estimating peak RPM, and knocking off 400 rpm to allow for the heavy stock pistons

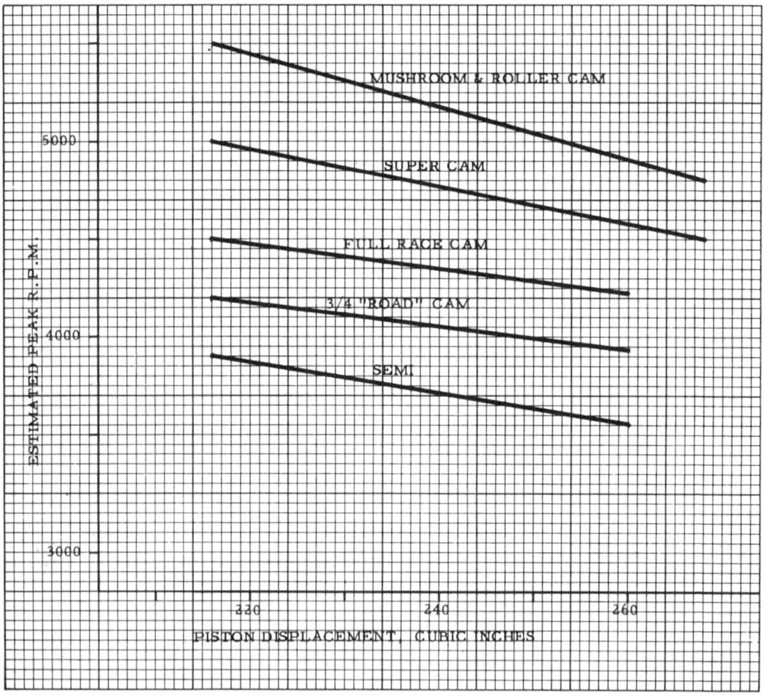


*Fig. 9-2. Chart for estimating peak RPM of a souped Chevrolet engine (aluminum pistons)*

(cast iron), we get a peak RPM estimate of about 3800. Therefore our estimated peak is 131 hp at 3800 rpm.

## ESTIMATING PEAK TORQUE

This part of the job may be a little more complicated, but it will be just as easy to work out and we can get good accuracy on it. You will recall that the peak of the torque curve comes at something around ½ the peak RPM and that the peak torque was some 25-30% above the torque at peak HP.

When we soup a stock engine, we juggle these relationships somewhat, but the principle is still the same. We can take advantage of these facts to calculate the HP at peak torque without ever actually touching a torque figure. Here's how:

If torque remained constant at all speeds, obviously HP would fall off in direct proportion to the RPM. However, since torque is rising as we reduce speed, the HP will not drop in proportion and we must work out a certain multiplying ratio in order to get the peak torque point on our graph. Use this formula:

$$HPt = \frac{RPMt \times R \times HPp}{RPMp}$$

where HP*t* is the HP at peak torque; RPM*t* is the RPM at peak torque; HP*p* is the peak HP; RPM*p* is the peak RPM; and R is the peak torque ratio. Below we have listed approximate torque ratios ("R") and peak torque speeds (RPM*t*) for the various cam grinds on the Chevy block:

TABLE II

PEAK TORQUE RATIOS

| | |
|---|---|
| Semi | 1.22 @ 2000 rpm |
| ¾ | 1.20 @ 2500 rpm |
| Full-race | 1.18 @ 2800 rpm |
| Super | 1.15 @ 3000 rpm |
| Mush., roller | 1.10 @ 4000 rpm |

As an example, suppose our estimated peak is 160 hp at 4600 rpm, and we are running a full-race cam. Then our estimated peak torque point would be:

$$HPt = \frac{2800 \times 1.18 \times 160}{4600} = 115 \text{ h.p. at } 2800 \text{ rpm}$$

With this and the peak HP value, we can lay out an accurate power curve for a souped Chevy.

## WORKING OUT AN EXAMPLE

Suppose we have a late standard 90-hp block and we add the following "soup": Milled and filled head, 9:1 compression; ported and fitted with large valves; triple manifold; super-race cam; 1/16 bore and Powerglide crankshaft; special ignition and lubrication. The problem is to estimate the full power curve.

In the first place, the displacement would be 235 cu. in., which is a 9% increase, so the increment (from Table I) would be 0.9 x 9 = 8 hp.

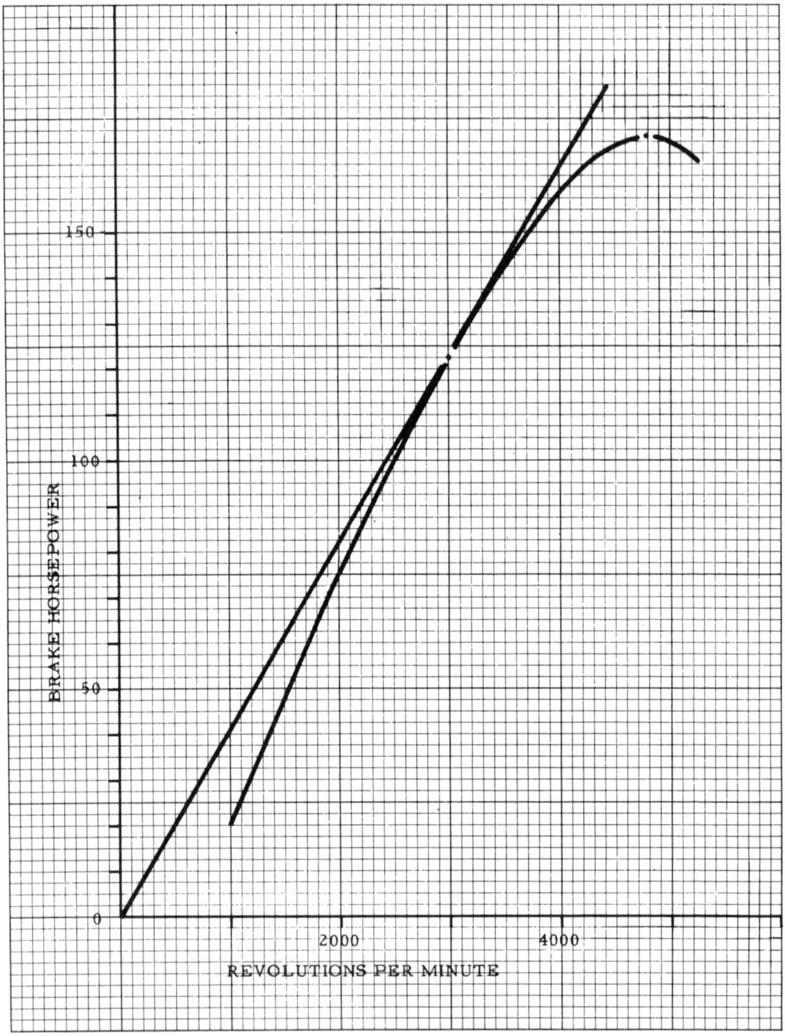


*Fig. 9-3. Estimation of H.P. curve for the example in the text.*

Other increments: Compression ratio (from Fig. 5-2 Chap. 5), 14; porting, 4; large valves, 5; manifold, 17; cam, 30; miscellaneous, 3. Adding them all up gives 81, so the estimated peak HP is 171. Using Fig. 9-2 gives a peak RPM estimate of 4800. For estimating peak torque, we use the formula and torque ratios in Table II thus:

$$HPt = \frac{3000 \text{ x } 1.15 \text{ x } 171}{4800} = 123 \text{ h.p. at } 3000 \text{ rpm}$$

The rest is simple. Fig. 9-3 shows how we construct the final power curve from the above estimated figures, according to the rules laid down earlier. We realize that all this will be an hour's job if you can't use a slide rule, and we can understand your giving it up as wasted effort before you finish. But we feel the results are well worth while. We strongly advise that you estimate a power curve if you're setting up a fast chassis, especially a track job. (Incidentally, we haven't listed HP variations when burning fuels other than gasoline; refer back to the fuel section of Chapter 6 for further data.)

## MORE ABOUT SOUPED PERFORMANCE

It might help inexperienced rodders to get a somewhat sharper idea of the true HP potentialities of the Chevy engine if we quoted a few actual figures and looked at a few dynamometer curves. It is vital to the success of your work, and to the satisfaction you get out of it, that you know what power you are getting — and what you can and *can't* get.

Your stock Chevy engine as it stands, has no high RPM potentialities. Those heavy, cast-iron pistons see to that. By replacing these with aluminum pistons, you boost the peak speed by 300-500 rpm right off the bat! Using the stock iron pistons, we doubt that you could ever get a peak much above 4000 rpm with all the souping tricks in the world.

Otherwise, when you drop in the light pistons, RPM figures for the Chevy get right into a more normal range — that is, about 3800-4200 rpm for mild road engines; 4500-5000 rpm for the track jobs, with a potential of 5500 for the Wayne head and radical cam timings. You will notice that these figures correspond closely with those for the souped Ford-Merc V8. The fact that we are equalling V8 speeds while working with considerably larger bores speaks well for the basic layout of the Chev.

As for the HP itself, the Chevy takes no back seat either. For an inexpensive road engine (Powerglide block) — that is, assuming we stick with the stock head, rods, lubrication, etc. — and limiting ourselves to 8:1 compression on pump gas, the power potentiality would be around 160 hp at 4300 rpm. (This would be 150 hp at 4400 rpm with standard block.) Fig. 9-3 shows dynamometer curves for a mild road setup by McGurk. If we want to shoot the works on a Wayne road conversion, we could realize about 190 hp at 4300 rpm on pump gas.

For strict competition work, it is assumed that we can spend some money and with the possibility of using huge displacements, super compression, radical cams, and methanol fuel, HP horizons extend considerably. The very maximum piston displacement with the Chev block will be about 270 cu. in. (with sleeving). At this displacement, we would estimate your maximum potential output with stock head reworked (12:1 compression), but with a full house otherwise, as around 220 hp at 4300 rpm on methanol.

Wayne has achieved 263 hp at 4500 rpm with Hilborn fuel injection and methanol; Howard's have bled 267 hp from this displacement with their own fuel injection and roller cam. These outputs are virtually 1.0 hp/cu. in. — which slightly outdoes the V8 with overhead valves in this size range. Wayne's competition conversion for the SCTA 250-cu. in. Class B has been run to 244 hp at 4700 rpm, with 10½:1 compression and Hilborn fuel injection (248 cu. in.). Fig. 9-4 shows additional power curves for this engine. The small Wayne-Chevy for the AAA "sprint" class (220 cu. in.) develops 216 hp at 4900 rpm.

So it is obvious that the Chevy block will put out very nearly 1.0 hp/cu. in. on straight methanol fuel right down the line, if it's built up right. How much higher we might go without resorting to super fuels is speculative. They get 1.3 hp/cu. in. at 5500 rpm with the well-known

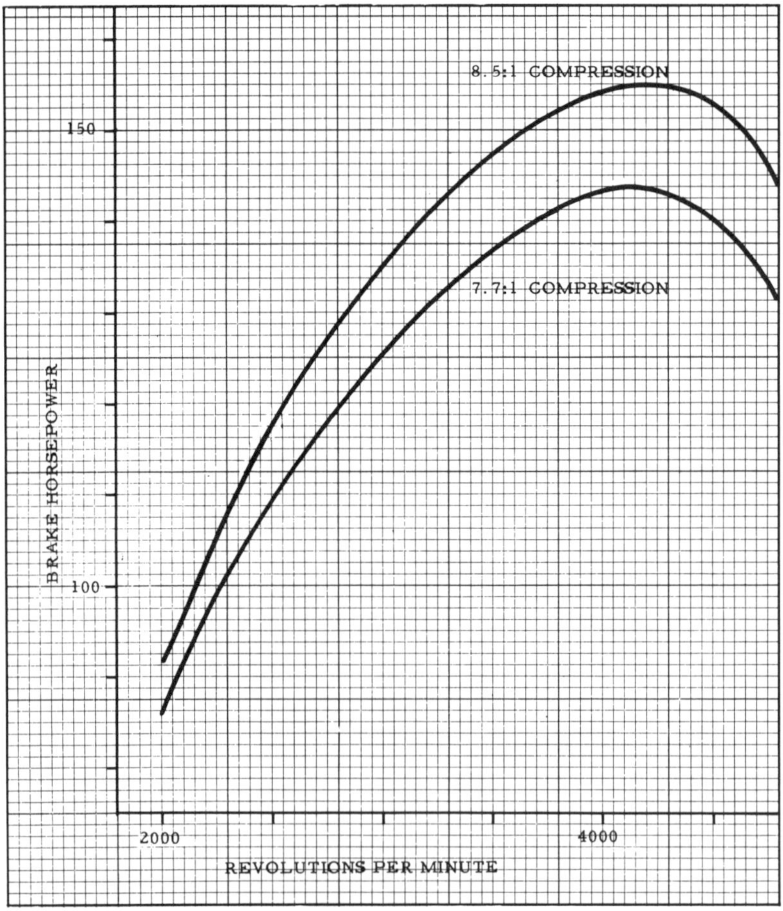


*Fig. 9-4. Dyanometer test curves for McGurk-Chevrolet road engine.*

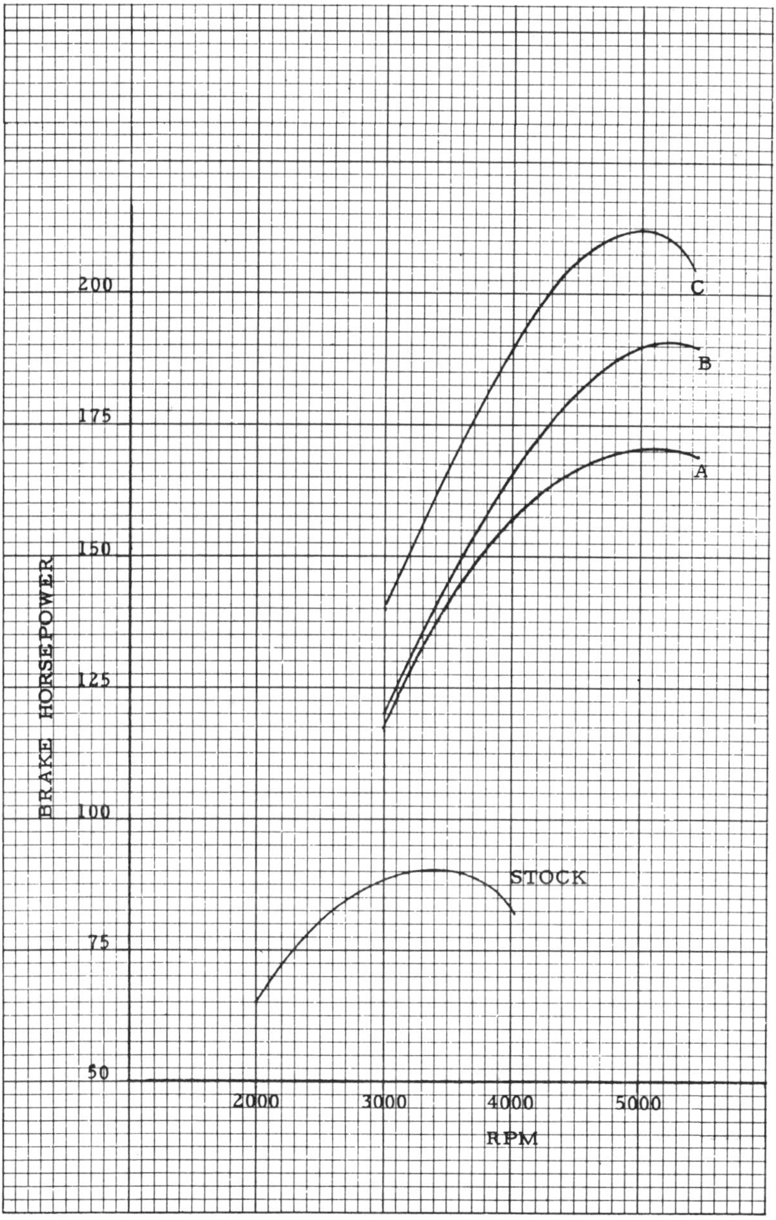


*Fig. 9-5. Some power curves for the Wayne-Chevrolet conversion.*
A. *Full Wayne conversion; 248 cu. in.; 10½:1 compression; single carb; 91/96 octane gas.*
B. *As above, but with three carbs.*
C. *As above, but pure methanol fuel.*

*Fig. 9-6. Intake side of a Wayne-Chevrolet marine racing engine—248 cu. in.; 12:1 compression; Vacturi carbs. This setup will develop around 229 hp on methanol.*

*Fig. 9-7. Exhaust side of the marine racing installation shown in Fig. 9-6. The large fuel pump returns overflow fuel from the floatless Vacturi carbs to main fuel tank, and also maintains pressure on the main tank to feed fuel to carbs; round tank in front is for the dry-sump lubrication system.*

"Offy" 4-cylinder racing engine without too much trouble. But we can never do that with the Chevy until we get a head with huge *inclined* valves. Even then, we could not likely sustain 350 hp with a 270-cu. in. stock block — but this subject at least points up the fact that there is a *definite place* in the Chevrolet souping field for a cylinder head with inclined overhead valves, operated either by pushrods or double-overhead camshafts.

In concluding this chapter on performance, we point out once again that, for a road engine, you must resign yourself to the HP limitations imposed by pump gas, muffling, and cam timing. These are "power killers" that you have to live with. So don't throw convention to the winds and go off the deep end on these items — and then find yourself with a car you can't enjoy. Oh, you can get a super-race mill around town, but the unending wrestling match gets a little inconvenient after three months! So take our advice: With a road engine, don't go above 8½:1 compression, and don't go over a ¾ cam under 230 cu. in., or over a full-race under any conditions.

So now we have finished the souping job and we know what the engine will put to the clutch. Now let's get to the *big* item — planning the job. Here is where the soupers tear hair; each one has a different idea of exactly how to proceed. So we will let our trusty old slide rule call the signals. (They say slide rules don't lie — but that liars can work slide rules!)

*Fig. 10-1. A Wayne-Chevrolet engine prepared for South American racing; this one has a sump lubrication system.*

*Fig. 10-2. Wayne track engine with two Riley side-draft Indianapolis-type carburetors.*

# CHAPTER 10

## PLANNING THE JOB

YOU heard about the Leadfoot Louie who souped his Chevy without planning. The first thing he did was to order a reworked head of 8½:1 compression, the maximum he figured he could use with pump gas. So what happens when he bores 3/32 in. and sticks in a Powerglide crank? His displacement shoots up to 239 cu. in., his compression ratio to 9.3:1—and he's got a bad case of knock on his hands.

He figures if he can cut off a little of the breathing at low RPM, he can get rid of the knock – so in goes a super-race cam! That eliminated most of the knock, but then he had to hold his rod in second gear to keep it running around town. He tried to remedy this by opening up the tappet clearances 0.050 in.; this helped, but his Chevy began sounding something like a cement mixer.

Our hero is now hanging around race tracks trying to peddle the whole kaboodle.

### FIRST THINGS FIRST

Intelligent planning is without doubt the most important factor in a good souping job. This includes looking to the future as well as today. When you begin to see dual manifolds in your dreams, you have the bug – and you'd best begin to *plan.*

The very first thing you must think about is just how much money you can sink in the job right away and, if possible, estimate how far you can go in the foreseeable future. This is not easy, goodness knows. "Souping" is a germ that grows on you; the more "guts" you pack in that engine the more you are going to want. (In fact, we're beginning to think that the thrill derived from violent acceleration and speed is as natural to the human male as hunger.)

If you are a married man with pressing financial responsibilities, and there are a lot of them in souping work these days, you may not be able to set your sights very high. That's all right – you can do wonders with $50 – and who are we to deny Junior a new pair of shoes? But the whole point is this: Don't just act on a brainstorm, wander into the nearest speed shop, and buy a dual manifold. If you want to do some souping, sit down and think a minute. Decide how much money you can afford to spend *immediately,* then try to estimate future possibilities.

If, for one reason or another, you can't do any of the actual work yourself (for reasons of lack of tools, time, or know-how), a good time to start your souping is when you have to have your engine torn down for an overhaul. You can then buy the special parts and have the mechanic put them in when he assembles the engine. Always keep little things like this in mind.

(And incidentally, there is a favorable time element involved with the novice souper. If you live in the northern half of the country, start your souping in the spring or summer; it takes an old hand to appreciate a hot engine at 10 below zero!) At any rate, we want to emphasize this need for definite financial planning before you ever lay wrench to bolt in this souping business. After that, the job is merely one of selecting the equipment and methods that will best fill performance requirements within cost limits. This is no small job.

## COSTS

It would probably be best to start off with a summary of approximate costs for the major items of speed equipment and related operations as applied to the Chevrolet engine. Here is a rough guide with which to estimate costs:

| | |
|---|---|
| Labor cost for pulling, tearing down, and reassembling Chevy engine | $70 |
| Semi-reworked stock road head | 50 |
| Full-reworked stock head | 70 |
| Milling head and sinking valves | 30 |
| Compound manifold | 25-45 |
| Carburetors | 15 |
| High-lift rocker arms | 12 |
| Boring | 15 |
| Powerglide crankshaft | 37 |
| Aluminum road pistons | 25 |
| Aluminum racing pistons | 40 |
| Special insert-type rods | 30 |
| GMC rods | 30 |
| Drilling Chevy crank for pressure oiling | 25 |
| GMC oil pump | 14 |
| Chopping flywheel | 10 |
| Regrinding camshaft | 30-40 |
| Tubular pushrods | 10 |
| Special valve springs | 4 |
| Large intake valves | 10 |
| High-output coil | 8-12 |
| Double-breaker ignition conversion | 7 |
| Complete double-breaker ignitions | 15-25 |
| Complete dual breaker-coil ignitions | 50 |
| Magnetos | 60-110 |

While we're on this subject of costs, it might be interesting to list the major souping methods in the order of their average boost in peak HP per dollar of cost. In other words, if a reground cam cost $30 and increased peak HP by 20, our power boost would be costing us $1.50 per HP; this would be contrasted with say $4.00 per HP by using a dual manifold. Here's a listing of the major souping methods on the standard block, based only on *peak* HP (the most economical method is at the top):

1. Reground cam
2. Compound manifold
3. Reworked head
4. Increased displacement

If we consider only the HP boost at low speeds, around 1000-2000 rpm, the "economy" listing is juggled as follows:

1. Reworked head

2. Increased displacement
3. Compound manifold (often decreases HP)
4. Reground cam (decreases HP)

Which of these lists you follow, of course, must depend on what performance you want from your car. If you are looking for an engine with worlds of power at low speed for scooting around in city traffic in second and high gear, the best bet is to increase compression ratio and piston displacement; if you want a hot car for highway travel, with all its "guts" above 50 mph, the most economical and effective souping paths are the cam and manifold.

## THE SOUPING CATEGORIES

Let's lay out some specific procedures for the amateur souper to follow. Probably the easiest way of doing this is to set up several what we might call "souping categories," based on cost limits. The cost limits will have to be more or less arbitrarily chosen, but we will try to select them to best suit average cases. It will also be necessary to sub-divide each category into two sections, since some men will be able to do most of their own work, and can therefore buy more speed equipment within the cost limit, through the saving on labor costs. With these facts in mind, then, let's establish five general souping categories with approximate *total* cost limits as follows (assuming you already have the basic Chevy engine to work with, and that it doesn't need a lot of extra work to put it in shape):

1. Light road engine—$50
2. Mild road engine—$125
3. Medium road engine—$250
4. Hot road or track engine—$500
5. Super competition engine—$1,000

From the above cost limits, you can probably fit your own case into one of the categories. Let's see what can be done in each one:

*Light Road Engine*—$50. Don't expect miracles for $50! This amount won't even cover much labor. However, there are certain things we can do. Since the resulting HP output on $50 will be pretty mild, we won't need to worry about a dual exhaust system, and can concentrate on the engine itself (though a $6 single Hollywood muffler should add about 5 hp at the top end).

Now for a total cost of around $50, there are really only two other possibilities, assuming you must hire the work done: (1) Milling the head and sinking the valve seats, and (2) a dual manifold with its extra carb. Milling the head is the best move for a city car; a compression ratio of about 7½:1 is good for this category (see Fig. 5-4, Chapter 5, for amount to mill off). The dual manifold will be best for the highway car.

If you can do your own engine work, or if you are having the engine torn down for other repairs, you could do either of the above two, or could install a ¾ reground camshaft; this latter would be the best move for the highway job, but it's not good for city driving. (We don't recommend a "semi" cam grind in this case.)

On a strict HP basis, the power increase with these steps will run about 8 hp for the mill job, 15 hp for the dual manifold, and 20 for the cam; peak RPM will be around 3400 for the first two, and 3800 for the cam. You pay your money and take the choice!

*Mild Road Engine* — $125. We have a lot more leeway with more than $100 to spend. But even now, don't begin thinking about staying with a Wayne-equipped car. We still won't consider a dual exhaust, though we definitely recommend a straight-through muffler at this point.

If you can't do your own work and there is no occasion to tear the engine down, we recommend you buy a commercial semi-reworked cylinder head (exchange) with 7.7:1 compression, and a dual manifold. This setup will give a good power increase over the full speed range and will yield a peak of about 120 hp at 3500 rpm on the standard block (about 135 hp on the Powerglide). If you can do your own work on assembly, etc., there are several possibilities; you could have the head milled and sunk to 7½:1 compression, plus a dual manifold and ¾ cam. This would give the best peak of around 135 hp at 3800 rpm, but would have little effect at low speed.

For better low-speed performance, you could buy a full-reworked head of 8:1 compression plus a dual manifold. If you are interested only in low-RPM output for exclusive city driving, the $125 could get you a 3/32-in. bore job, special aluminum road pistons, a Powerglide crankshaft, and a head mill job; this setup would give only around 110 hp at the top, but you would get terrific punch in the low-speed range. Stock ignition should be satisfactory in this category, though you might consider a special coil or inexpensive double-breaker conversion kit. Stock lubrication and bearings will be all right.

*Medium Road Engine* — $250. We're at the point now where that engine must come all the way down, one way or the other. This alone will represent about $70 labor cost to the boy who can't do it himself. That can buy a lot of equipment. Also we are at the point where we need to stick a few special parts in that engine, such as insert rods, just for the sake of reliability and durability at high output — parts that don't pack much HP! So keep this in mind in your "medium-road-engine" pipe dreams.

A dual exhaust system is probably not essential to the success of this engine, but we would definitely specify a straight-through muffler, and dual exhaust if possible. Otherwise, if you are in the "white collar" class— that is, if you can't do your own engine labor — we would recommend the following steps in this category (for the standard block): Milled and sunk head, 7½:1 ratio; dual manifold; ¾ reground cam; 3/32 or ⅛ bore; special aluminum road pistons; Powerglide crank; special insert-type rods; chopped flywheel; inexpensive ignition conversion (double-breaker).

You are probably wondering why we didn't include a commercially-reworked head in the above. Well, to stay within the $250 limit, we had to eliminate something, and since we consider insert rod bearings as *absolutely necessary* at this point, it seems most practical to add all parts that require engine disassembly. Later you could add a reworked head without tearing the block down. Consequently we have included a bore job, special pistons, and a Powerglide crank. Also, we've taken advantage of the engine "down time" to stick in a ¾ cam; to take full advantage of this requires a dual manifold, so the only thing left to omit was the head.

If you could do your own work, you would save the labor cost and could squeeze the reworked head in under the $250, or thereabouts. This would be a good idea. The main thing to remember is to plan the job so that all internal engine work that you can possibly foresee will be done at once; then you can add external parts later, as you can afford it. With the changes outlined above, our peak output would be 145-155 hp at 4100 rpm on the standard block, and we get good output at the low speeds.

Here is another point to keep in mind: We're now at the "crossroads" in souping on the Chevy engine, you might say. We have about reached the limit of the ability of the stock lubrication system to handle the job; further HP increases beg for 40-60 lbs. pressure to the rods and mains for maximum reliability and durability. So if you plan to go further than this "medium" category in the future, you had better do some very serious thinking at this point. Indeed, it might be better to wait until you can afford another couple of hundred and have the whole job done with GMC rods and oil pump.

*Hot Road or Track Engine* — $500. We can pretty much "cut loose" with $500 to spend. We couldn't squeeze a Wayne head in here along

*Fig. 10-3. Wayne-Chevrolet engine installation in the "Johnson Spl.", 220-cu. in. AAA sprint car; the engine develops over 200 hp.*

with everything else, but we can go all the way using the reworked stock head. A full dual exhaust system should be used here too. Otherwise, here are the modifications we suggest for your hot road or track engine:

1. Full-reworked stock head; large valves, ported, etc., 7½:1 rated compression ratio (10-12:1 for racing)
2. Triple manifold (possibly six carbs for racing)
3. Full-race cam (super cam for track)
4. Tubular pushrods
5. 3/32 or ⅛-in. bore
6. Aluminum pistons, (type depending on use)
7. Powerglide crank, drilled for pressure
8. GMC rods
9. Special bearings and bearing clearances (see Chap. 5)
10. Full pressure lubrication with GMC oil pump
11. Chopped flywheel
12. Dual breaker-coil or magneto ignition

The above combination should give you a peak of around 170 hp at 4300 rpm in road trim on pump gas, and fairly good torque at the lower speeds (though it won't outdo the stock setup below about 2000 rpm). You'll notice that we have specified a 7½:1 rated head; this is so when you allow for the increased displacement, the true ratio won't be too much above 8:1. We suggested a triple manifold here as it will give a bit more performance at high RPM; for city driving, a dual is better. A full-race cam will be all right here for city driving because of the large piston displacement we are working with; it should idle fairly decently and get around in traffic—but don't expect performance much above *stock* below 55 mph in high gear.

We have specified a complete conversion of the lubrication system for this HP range. The stock Chevy "splash-pressure" system is all right for low sustained outputs, but when we begin putting more than 150 hp through the rods and crank at speeds over 4000 rpm, things begin to come apart. It then becomes necessary to use thin, hard bearing linings and direct oil pressure of 40-60 lbs. to the rods and mains. This can easily be accomplished by drilling the crankshaft and using a GMC truck oil pump. We consider this lubrication system modification absolutely necessary for a satisfactory Chevy engine in the output range well over 150 hp. Not only this, but we could go on to a Wayne head later without major internal changes.

*Super Competition Engine*—Here is where we definitely "throw the book". At $1,000 you can afford about anything in the way of special equipment and engine work, and the output potentiality jumps to about 1.0 hp/cu. in. unblown. As things stand now in this souping field, the ultimate in Chevrolet competition equipment is the full Wayne conversion. This is our recommendation here. In other words, if you are after the last possible HP, and cost is no pressing object, *don't fool with a reworked stock head*! Definitely go to the Wayne head and related equipment. (Instructions for fitting up a Wayne competition conversion are given in Appendix I.)

Incidentally, if you are building up a Wayne conversion yourself, we suggest you follow these instructions pretty closely; they are the result of

*Fig. 10-4. Wayne engine installation in a track roadster chassis, running Hilborn-Travers fuel injection (the injector pump is driven from the front of the camshaft).*

a lot of grief on the racetrack — so benefit by bitter experience and use the recommended fitting procedure. On the other hand, if you don't want to build one of these engines yourself, the Wayne Manufacturing Co. will sell you a complete unit, to a large combination of specifications and ready to install, for around $1,000 or a bit more. For an output of nearly 1.0 hp/cu. in. with a brand new block, this is a pretty decent buy. All in all, the Wayne equipment makes a very practical racing engine in the $1,000 bracket.

## LOOKING AHEAD

This business of "looking ahead" is so very important to the success of your souping, especially if your funds are limited and you are souping on a small scale, that we feel it warrants a little more attention at this time. Let's take a typical example:

Here's Joe Smith who suddenly "gets the bug" — sees new beauty in the lines of his '42 Chevy coupe. He won't rest until he's restyled the body and souped the engine. The engine in the car now has 110,000 miles on it and is exhausted. Joe's business takes him on the road a lot, where time is money; he has every reason to want to cruise at 70 mph — and needs plenty of stuff *left* at this speed. Furthermore, Joe is no Rockefeller; he's getting married soon — he has to pinch pennies — and he hasn't the time nor the equipment to do any of his own work.

So here's the situation. What a spot for some smart planning! First, Joe figures he has to have a new engine, and that he could *eventually* sink

$200 above that in souping it. From here on, there is a *right* way — and a *wrong* way — to soup that Chevy.

What are Joe's performance requirements? He's not interested in a 100-mph top speed, nor is he interested in flashing top gear acceleration at 20 mph to impress the urban gentry — he just wants an engine that will loaf mile after mile at around 3000 rpm, and have very good acceleration in the 2500-3500 rpm band. Now right away, one money-saving idea presents itself: Since we must drop in a new engine, we can have the mechanic do the internal engine work at this time.

Now, what do we do? Well, first, the extra stroke length of a Powerglide crankshaft will do very little for torque or economy at high speeds, so we can forget that. On the other hand, for light and reliable running at a sustained 3000 rpm, Joe is definitely going to need aluminum pistons and insert-type rod bearings. (Stock lubrication will be okay.) With the oversize rings, it would cost around $30 to overbore the engine; this wouldn't do a lot for high-RPM output, so let's save money and forget this too. In the interests of acceleration only, the flywheel should be chopped.

A reground cam is going to be the key factor in this particular case. A full-race grind, giving peak torque at around 3000 rpm, would probably be best. But with stock displacement, the low-speed performance, cold starting, idling, etc., would be a downright nuisance on a business car. We'll have to settle for a ¾ grind, but we expect Joe to do a little investigating on comparative cam timings before selecting one (to select the hottest).

Now by this time, Joe's pocketbook is getting a bit thin; he must now choose between increased compression with a reworked head and dual carburetion. Actually there is no choice — the dual carbs are necessary to properly feed the increased breathing of the cam. So Joe will just have to run with his stock head till he can afford something better.

So you see, there is a right way and a wrong way to do every souping job. If Joe had bought a reworked stock head with the money he used for aluminum pistons and insert rods, where would he be? Probably at the side of the road with an air-conditioned crankcase! The pistons and rods don't add a lot of HP, but they are vital to the success of ***this particular*** souping job. If Joe had been souping for city driving, they wouldn't have been so important.

In other words, when you are planning your souping job, try to take *everything* into account. What type of driving will you be doing most? Do you want long mileage between overhauls? Is gas mileage important? Is cold starting important? In what RPM range do you want maximum HP or torque? Do you want to "drag" through the gears to very high RPM? Consider all these, and a hundred other things.

In this book we couldn't hope to cover every possible situation, so we have just tried to suggest what to watch out for. If funds are sharply limited, your planning job will be that much tougher — and you'll have just that much more to watch out for. Sometimes you must choose between one "good souping deal" and another. This sort of thing makes strong men weep. Or maybe you can eventually spend $500, but can only muster $50 in the next three months. What will you do?

We think it might be a good idea to close this book by reviewing general steps that are effective in several different "souping situations", to form a rough guide for the novice. They are:

*If you are souping lightly and want maximum economy:* Increased compression; dual carbs (with lean jets); semi cam.

*If you want a car that will be quick around town:* Increased compression; semi cam; bored out with Powerglide crank; dual carbs (with small venturis and rich jets); Roots-type supercharger.

*If you want a fast car for the open highway above 45 mph:* Ported; bored out; aluminum pistons; insert rods; triple carbs (with large venturis); full-race cam; centrifugal supercharger.

*For an engine with maximum HP at all RPM:* The works!

That last isn't exactly a technical statement, but we think you get the idea. In fact, we were wondering if we should add a fifth category, for "the best all-around road engine". The Chevrolet front office would probably be happy to give us some advice in this direction—*leave it alone*! But as a psychiatrist would say, "That's taking the negative attitude." And besides, that is no way to act if you want to get that Chevy to go. It will go, too, if you plan properly.

In this book we have tried to steer you in this souping business. We have soft-pedalled specific instructions on machine work because we realize that not one in 25 will be doing such work; and for those who do, literature is widely available. We have concentrated instead on answering the questions on *planning.* We have tried to leave no major decision in this field up to the amateur without specific advice.

You must carry the ball from here. All we can say now is: *Use your head. Plan the job carefully before you loosen a bolt. Think of everything. Look ahead.* And remember the words of our moron hero, Leadfoot Louie, who speaks from bitter experience:

"No light, flyweight pistons for "Chevy" McBlow;
He stayed strictly stock—with no bugs;
But his only reward for five-thousand in low
Was a lapful of cast-iron slugs!"

*Latest Nicson dual-intake manifold for 1937-51 Chev. models. Note heat risers, Zenith carbs, special air cleaners.*

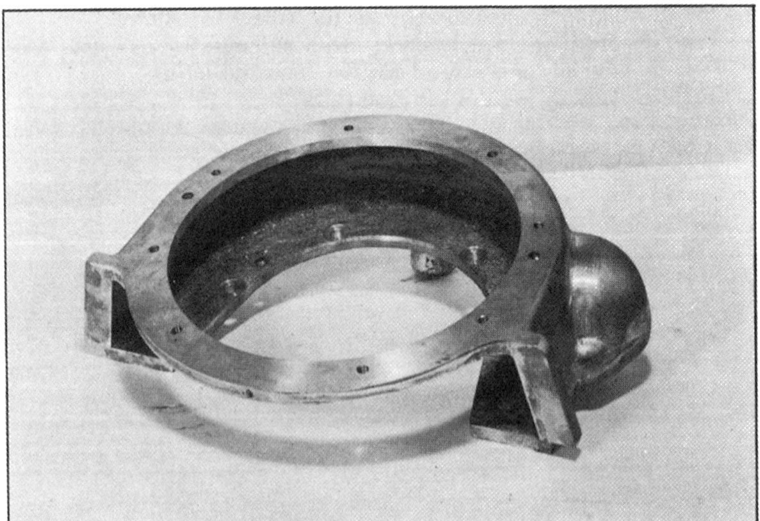

*Bell housing adaptor for installing Chev. or GMC engines to Ford transmission.*

# APPENDIX I

## CLEARANCES & GENERAL OPERATING DATA — WAYNE ENGINE

MAIN JOURNALS: McQuay-Norris bearing inserts recommended. For 1940 to 1947 blocks use following McQuay-Norris numbers: 2875 — 2876 — 2878 — 2994. For 1948 and later blocks use numbers 3528 — 3529 — 3530 — 3531. If crank is standard size (not ground) use .020 undersize inserts, sizeable to standard type. Line bore to .003 to .0035 total clearance on crankshaft.

Note: If crankshaft has been ground .010 under standard, use the .030 undersize inserts, sizeable to standard type. The number three main (thrust) should be faced off to give .005 thrust clearance.

MAIN BEARING CAPS: Block should be smooth so that caps do not "rock". Do not use shims under caps. Lock washers are not necessary. Main cap bolts should be tightened to a torque-wrench reading of 110 foot pounds. Check all main bolts to see that they do not "bottom" in holes, shorten if necessary. Tap two 5/16 inch holes in number one front main cap to ⅜ inch N.C.

CONNECTING RODS: Use G.M.C. 228 truck rod, part #2135418. McQuay-Norris bearing inserts #2980, .020 undersize, sizeable to standard recommended for standard size crank. If crank is .010 under use the .030 undersize insert. Have rods line bored or sized to have .004 to .0045 total clearance on crankshaft. Close the small oil hole in side of rod that sprays oil on cylinder wall. Use a punch to peen hole shut.

Note: The G.M.C. rod when used on the 1940-47 cranks will have about .060 side play, this is permissible and has no ill effects.

PISTONS: Wayne-Venolia pistons are finished to have proper clearance to fit bored size for which you order. Pistons are finished .002 larger on skirt diameter, and are ground to fit cylinder with .010 clearance just below rings, and .008 clearance at skirt.

RINGS: Pistons can be furnished with ring grooves for 1/16 compression, 1/16 scraper, and 5/32 oil ring, or 3/32 compression, 3/32 scraper and 5/32 oil ring. Ring gap should be .015 to .020.

BLOCK: If block is being prepared for installation of Wayne head, grind or file side of block to provide clearance for the large ball end of the push rod when inserted thru push rod hole in cylinder head. Starting at top of block, directly above each valve lifter hole, file a tapered groove 1/16 inch deep, ½ inch wide, and one inch long down side of block.

Drill additional crankcase breathing holes between each valve lifter hole, to allow crankcase pressure to be relieved into push rod chamber. Use a 5/16 drill, do not drill into No. 2 and No. 3 main bearing webs.

Remove sheet metal end plugs in main oil passage. Tap with ⅜ pipe tap, holes are proper size to take tap, no drilling is necessary. Be sure to tap hole at rear deep enough, so that pipe plug will go flush and not interfere with bell housing or rear motor plate.

Enlarge two lower bolt holes of timing gear cover and front motor plate to 13/32 diameter. Tap matching holes in No. 1 front main cap with

⅜ N.C. tap. This will permit removal of front timing cover and camshaft without necessity of dropping oil pan, as is the case with stock set-up.

Remove from block the sheet metal tube that normally supplies oil to oil spray tubes in stock pan. Enlarge hole to 37/64 and tap with ⅜ N.P.T. Connect G.M.C. 228 oil pump to this hole, using special Wayne flexible line and fittings. Use ¼ inch pipe plug in hole to which stock pump is normally connected. This will force all oil from pump to be passed out of block to Full-Flo (Chrysler Six—1947) Filter and then returned to block for distribution to bearings thru main oil gallery.

Note: If no filter is used a short "jumper" line in place of filter can be used to return oil to main oil gallery.

WRIST PINS: Wrist pins are G.M.C. 248 truck, .990 diameter. Part number is #6107357. Pins are full floating type and require aluminum plug or end buttons. We recommend that you let us furnish and fit pins and buttons with pistons. Pin with buttons installed should have .015-.020 cylinder wall clearance. Pins should be push fit in piston and rod bushing.

CAMSHAFT: Factory tolerances on cam bushings. Thrust plate should have .003 to .005 clearance between gear face. Valve clearances should be those recommended for particular cam grind being used.

CAMSHAFT GEAR: Aluminum gear set recommended. Consists of matched steel crankshaft gear and aluminum cam gear.

OIL PUMP: G.M.C. 228—part #2136436. Factory by-pass setting of 60 pounds is recommended. Use special Wayne flexible high pressure ½ I.D. line to connect pump to block.

Note: Remove oil tube in block and drill with 37/64 drill, tap ⅜ N.P.T., plug old oil pump line hole with ¼ pipe plug. Increase length of pick-up tube on base of pump one inch. Braze a one inch wide sheet metal band to oil screen so that screen will clear lengthened pick-up tube.

OIL PAN: Remove all oil troughs and lines. Cut out section to clear oil pump screen and flexible oil line. Use rear half of an old oil pan and weld to front section of cut out pan. Drill six ½ inch drain holes in shallow or foreward section of original pan. Add a horizontal baffle in rear section of pan, drill six ½ inch drain holes in it also. Place horizontal baffles (similar to stock pan) just forward and aft of oil pump screen. Bump out side and bottom of pan to clear pump body and screen. Note: It is advisable to set up pump and oil line in block and to do all cutting and forming for clearance before any welding is done. A pan reworked in this manner will hold 2½ gallons of oil.

IGNITION: Dual Coil—3 Lobe—Two Point System Recommended.

This ignition will jump a ½ inch air gap at 8000 R.P.M. The automatic advance is at a rate of 10 degrees per 1000 R.P.M.'s and is set at 5 degrees B.T.C. for preliminary tryout. Depending on fuel and compression ratio being used the spark may be set somewhere between T.D.C. and 13 degrees advance.

COILS: Bosch TC 606—Big Brute coils are recommended. Voltage output is about 25000 volts.

## INSTALLATION WAYNE CYLINDER HEAD

1. On some cylinder blocks it is necessary to provide clearance on the block for the large ball end of the push rod when inserted thru push rod hole in cylinder head. Starting at top of block, directly above each valve

lifter hole in block, file or grind a tapered groove 1/16 inch deep, ½ inch wide and one inch long down side of block.

2. Remove rocker arm assembly, inserting 6 — ⅜ cap screws in aluminum rocker stands as you lift assembly off of long studs. This will hold assembly together and facilitate reassembly.

3. Cut the heads off of two long stock head bolts. Screw these into block on same side as intake ports of head. These will serve as guide pins, holding gasket in place and aiding in positioning head.

4. Be sure all head bolt holes in block are clean — do not have any liquid in holes.

5. Place head in position on block. Before inserting head bolts, lightly oil threads and head bolt washers. Tighten bolts starting at center and working out. Use ⅜ socket on cut off section of Allen-wrench. Tighten bolts progressively up to a torque-wrench reading of 85 foot pounds.

6. Install push rods — install rocker arm assembly — back off all rocker arm adjusting screws before tightening rocker arm assembly to head. Connect rocker arm oil line.

On full pressure jobs running 60 pounds oil pressure, solder up fitting that brings oil out of block to rocker arm oil line. Redrill soldered up fitting with a No. 60 or No. 70 drill. This will tend to reduce oil pressure at rocker assembly.

7. Adjust tappets per clearances furnished with camshaft.
8. Install intake manifold.
9. Install valve cover — shellac gasket to head only. Apply about a ½ pint of oil to valve mechanism before starting engine.
10. Install side plave — trim head gasket if it interferes with side plate gasket.
11. Install exhaust manifolds. Note: It is not necessary to use exhaust gaskets as both surfaces are machined.
12. Run engine at partial throttle until thoroughly warm.
13. Remove valve cover and intake manifold. Remove two piece oil fitting on rocker shaft. Retighten head bolts, starting at center, to a torque-wrench reading of 90 foot pounds.
14. Readjust valve clearances.
15. Reassemble.

## GENERAL OPERATING DATA FOR WAYNE COMPETITION ENGINE

LUBRICATION: S.A.E. 40 or 50 Valvoline oil recommended. Pressure should not drop below 40 pounds when hot. Oil pan capacity 10 quarts.

FUEL: 10:1 Compression ratio. 1—91 octane aviation gas with 20% Benzole added. Percentage of Benzole may be increased to 40% if necessary. 2—Methanol (alcohol.) No additives required. 12:1 Compression ratio. 1—Methanol (alcohol.) 14:1 Compression ratio. 1—Methanol (alcohol.)

SPARK PLUGS: Champion 14 MM, J-2 or J-3 series recommended.

SPARK TIMING: Set timing on T.D.C. for first run. Increase by 5 degree steps up to 10 degrees. Final setting may be somewhere between 10 and 15 degrees. Distributor should have 10 degrees advance for every 1000 R.P.M. of engine speed.

CAMSHAFT: Check valve mechanism regularly. Valve clearance should be held to close tolerances for good performance. Inspect cam followers for excessive wear if valve clearances continue to increase.

VALVE SPRING TENSION: With valve on seat, overall length of spring should be 1-13/16 to 1-27/32 inches. If valves are ground or seats refaced add spacer washers to bring spring length back to dimensions given.

*Souped-up GMC 270 conversion mounted in special chassis. Car has run in several hot rod classifications, will do 150 Mph.*

*Owner of this car is specialist on GMC speed equipment, Howard Johansen. Note 5-carb manifold.*

# APPENDIX II

## SOUPING G.M.C. ENGINES FOR CHEVROLET CARS

Just recently, a new basic stock engine has flashed upon the American souping scene — the small GMC truck series. With relatively minor modifications, these blocks can be installed in the Chevrolet chassis to provide displacements up to 300 cu. in. and outputs of over 200 hp on pump gas!

The main advantages of these engines over the stock Chevy are: (1) Possibility of greater piston displacement and HP, (2) lighter weight of only 525 lbs., (3) overhead valves, (4) an extremely rugged lower end with four large copper-lead main bearings, short stroke, very stiff, short rods, and (5) full pressure lubrication at 50 lbs.

It's an ideal setup for souping purposes; it will "stay put"—and it will *go*! Prices are a bit higher than when working with the Chevy block, but since the lower end will pump fantastic torques at high RPM without major changes, costs level out as the degree of souping goes up. Let's take a look at this wonderful engine:

### GENERAL

We're concerned here only with what they call "Group I", or the small series of GMC truck blocks, which includes the models 228, 248, and 270 (the model numbers designate piston displacement). The 302 and larger models are too bulky for our purposes.

All these models use the same basic block, with slight variations in cylinder wall thickness and block height to allow for different bores and strokes. Cylinder heads, crankshafts, rods, and bearings are interchangeable. The 228 and 248 models have a stroke of 3-13/16 in.; the 270 has 4 in. Bores are 3-9/16 in. on the 228, 3-23/32 on the 248, and 3-25/32 in. on the 270. Otherwise, the three engines are very similar.

Actually, unless your piston displacement is limited in some competition class, your best bet is to work with the 270 block with its much greater piston displacement, since you will likely have to buy the engine separately anyway. The 270 is definitely the popular block today. Souping procedure closely parallels that for our basic Chevy engine.

### FITTING UP THE BLOCK

Very little needs be done with the block itself. For increasing piston displacement, bore it out and/or use the 4-in. 270 crankshaft ("stroking" is not a factor). The 270 block can be overbored 7/32 in., but the usual practice is boring a more conservative 5/32 to get 292 cu. in. The 228 and 248 blocks can be bored about 3/16 over.

Pistons for the various bores and strokes are not yet widely available. Frank Venolia supplies solid-skirt racing pistons; Horning, McGurk, and J.E. have racing types available also. Harry Warner, who bought out the Wayne Manufacturing Co., has come up with a fine idea. His new GMC head sets flat across the top of the block so that the combustion chamber is within the cylinder walls; a very short piston is used which weighs ⅓ less than any of the other types. (Then, of course, if you don't overbore,

*Vinola racing pistons for GMC engine available up to 3 15/16" bore. These pistons, used with standard GMC heads, produce compression ratio of 82:1.*

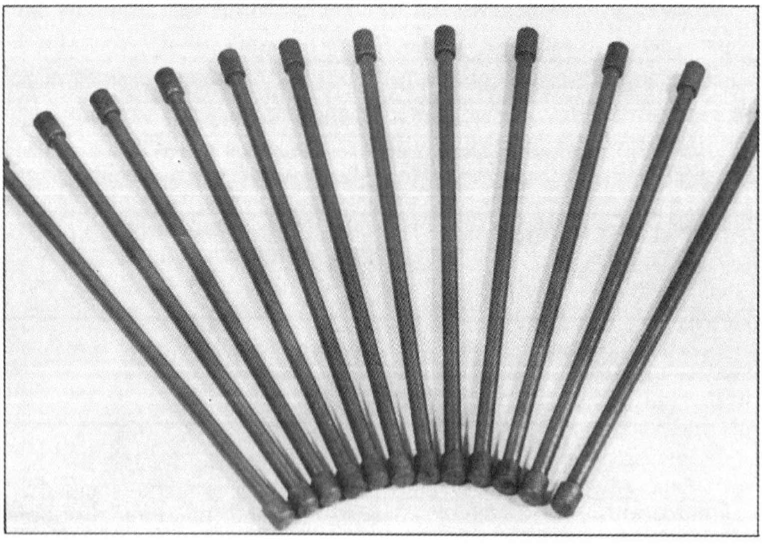

*Light-weight tubular push-rods for all Chevrolets, 1937-51 and GMC engines.*

the stock aluminum GMC pistons will work very well under high-output conditions.)

The lower end will require very little attention even under extreme conditions. Bearing clearance recommendations are about the same as for the Chevrolet engine (see Chapter 5). The stock insert-type rods will be okay without changes. For racing at high RPM, the crankshaft main bearing journals should be "grooved" about 1/8 in. to increase the oil flow to the rods. The lubrication system itself should require no changes. For racing, however, you may want more than the stock 50 lbs. oil pressure; a heavier relief valve spring can be purchased to boost pressure to around 80 lbs. Stock copper-lead bearing inserts will be satisfactory for most conditions; Federal-Mogul or McQuay-Norris cadmium-base linings are often preferred for competition.

## THE CYLINDER HEAD

For 1950, General Motors brought out an entirely new set of cylinder heads for the small block series, with much larger valves and ports, smoother induction flow, and an improved combustion chamber shape. The stock 270 head now carries 1.86-in. intake valves (compared with 1.64 on the old head), and the new head *alone* was responsible for boosting peak output on 6.8:1 compression from 104 hp at 3000 rpm to 120 hp at 3600 rpm! (The current 270 has 7.2:1 compression and is rated 130 hp at 3600 rpm.) The new 228 and 248 heads have smaller valves, and since all heads are interchangeable, it is definitely recommended that you work with the 1951 270-head with the high compression ratio.

With this new head, porting and increasing the valve port diameter are not necessary, though it is always a good idea to smooth and polish up the ports. As on the Chevy, however, the usual practice is to open out the intake and exhaust valve ports enough to narrow the seat to about 1/16 in. at the outside edge of the valve. (This is done to improve valve sealing as well as to increase the breathing, but is *not* an essential operation.) Larger intake valves can be fitted in these heads, but with the large port area of the 1950 and later models, it is not considered worth the bother. If you desire, though, you can fit 1.94-in. Chevrolet Powerglide intake valves in the late 270 head, and late 270 intake valves in the earlier heads.

For increasing compression ratio over stock, the GMC heads can be milled up to 1/8 in., or special domed pistons are available that increase compression by extending into the head. An 0.080-in. mill will raise compression to around 7½:1 on pre-1951 heads; compressions up to 10:1 are obtainable with domed pistons, more when coupled with milling. Filling is not generally practiced with GMC heads, as it kills off much of the performance attributable to the fine combustion chamber shape.

## SPECIAL HEADS

There are now several special cylinder heads for the small GMC block on the market. Wayne Horning has one available with machined, round combustion chambers, vertical valves, ample water cooling spaces, and six ports on each side. Most of the wearing parts are stock, to ease maintenance costs. Intake valves are modified Chevy Powerglide, exhausts are '48 Cadillac, guides and outer springs are Chevy, etc. To simplify the head porting layout, Horning has altered the valve sequence so that a special billet camshaft must be used. This puts the price tag up on the overall installation. It's a terrific performer, though,, and as high as 285 hp at 5000 rpm has been achieved on the dynamometer – from 274 cu. in.

As mentioned earlier, Harry Warner is working on a "Wayne" head for the GMC that fits flat over the block face, similar to the early 4-cylinder Chevy. This simplified layout, since it also utilizes the stock valve cover and side plate, will greatly reduce costs and still retain good performance.

Besides these two special heads, several companies supply reworked stock GMC heads on an exchange basis. These are milled, ported, and have oversize intake valves fitted. They are a very good buy for the boy who doesn't want to go too far with his GMC souping.

Let us remind you once again that the 1950 and later heads, with the large valves and ports, are definitely the ones to work with.

## SPRINGS, ROCKERS, PUSH-RODS, TAPPETS

With a reground camshaft, valve action is much more violent and we will need more spring tension to prevent floating at high RPM. Special springs are available from several companies to fit inside the stock spring to form a dual combination. These should be used with any reground cam; either stock 270 or Chevrolet springs can be used for the outer coil. The length of the springs with the valve on its seat should not exceed about 1-13/16 in.; if it does, shim up.

The stock GMC spring retainers are constructed of thin pressed steel; at high RPM you are apt to pull a keeper right through the retainer and "swallow a valve"–which is a good way to wreck an engine in the least possible time! Special retainers are available, and we recommend these for any competition. The stock keepers will be fine.

With the new GMC head in 1950 came a new rocker arm formed from two steel stampings welded together; this new rocker is much stronger and lighter than the earlier cast iron type, and should be used in racing engines. Stock GMC pushrods are solid steel, as on the Chevy. These are all right for a moderate amount of souping, but for high-RPM track racing, you need the lighter, stiffer tubular type; these are available from several outfits (they are interchangeable with Chevy pushrods). Stock tappets will be adequate for most conditions. For racing, many prefer the light '37 Chevrolet cup-type tappet; this can be used without changes with long pushrods.

## CAMSHAFTS

A reground cam is a "must" for almost all souping conditions. Many shops now grind GMC cams for very reasonable prices and to the usual timing ranges. Grind recommendations made earlier in this book for the Chevy engine can be applied here – that is, a ¾ grind for most road applications, and a full-race or super for the hottest road types with displacements near 300 cu. in. For competition work very good results are obtained with the Iskenderian, Herbert, and Howard roller and mushroom tappet setups. This deal is especially attractive with the Horning head, since it requires a billet cam anyway. Using these cams with port fuel injection, *over* 1.0 hp/cu. in. has been obtained on straight methanol!

## CARBS AND MANIFOLDS

Special dual and triple manifolds are now available from several companies to fit the GMC heads, at prices near that for Chevy manifolds (regular Chevy manifolds will not fit the GMC head). In addition, companies supplying special heads also have dual, triple, or six-carb manifolds available for their particular head casting.

As for carbs, a single-barrel 1.38-in. Zenith is standard for the 228 and 248 engines; a 1-in. duplex Holley is standard on the 270. These carbs are seldom used for souping purposes, and the same units that are popular for Chevrolet work (Stromberg BXOV-2, Zenith 28, Carters, Rochesters, etc.) are also widely used. Venturi diameters and jet sizes should be selected proportionally larger because of the greater displacement.

## IGNITION

All special ignitions for the Chevrolet engine will fit the GMC block without changes, and we can apply ignition recommendations directly from Chapter 7. As a rule, a converted double-breaker single-coil outfit will be sufficient for most all road applications, with the dual breaker-coil layout or magneto recommended for racing. Standard spark plug for the small GMC series is the AC "44" (14-mm). This will be all right for the road, or you could go a step cooler to "43". For racing, the "42" is best to start with; in terms of the Champion plug heat range, plug selection would run from the J-7 to the J-2 for racing. (See the tables at the end of Chapter 7 for others.)

## PLANNING THE JOB

We can't help but repeat that the basic GMC 270 engine is a well-nigh *perfect* stock souping unit. It has a rugged lower end that needs no major modification, a beautiful overhead-valve layout with huge ports and valves, and the whole outfit as it comes in the crate without manifold or clutch, weighs a mere 440 pounds. Unbelievable in terms of a truck engine—but there it is.

A neat and economical souping plan suggests itself immediately. Using a 1951 engine with the stock 7.2:1 compression ratio, bore the block 5/32 in. oversize and fit a set of special aluminum pistons. This raises the piston displacement to 292 cu. in. and automatically boosts compression to 7.7:1, which is a good ratio for road work on pump gas (or you could use domed pistons to give anything up to 10:1). For this mild degree of souping, a dual manifold, ¾ cam grind, and a double-breaker ignition conversion are in order. No changes will be necessary on the lower end.

At this point, we have roughly $180 invested in parts, and the true power peak, as installed in the car with dual exhaust, should be a sizzling 185 hp at 4000 rpm.

You could go a bit farther with a road engine—that is, you could use a commercially-reworked stock head, ported and with large valves, around 9:1 compression, triple manifold, full-race cam, dual ignition, and larger bearing clearances. This would put the parts investment near $350 (not including labor) and boost the output on pump gas to around 210 hp at 4200 rpm. For strictly competition work, you can go all-out with the Wayne or Horning head, roller cam, methanol fuel, etc., and get very near 300 hp at 5000 rpm on an investment of well under $1,000.

So you can easily see that we are getting a lot of "HP per dollar" with this GMC block. And at this time, we must conclude that the 270 model is the most economical and practical of all U. S. stock blocks for souping purposes. That's a broad statement, but the facts speak for themselves.

As time goes on, a much wider selection of special equipment will be placed on the market – more inexpensive road pistons, more special and reworked heads, more cam grinders, supercharger installations, etc. And about that time, the Chrysler V8 will probably begin running away from the field!

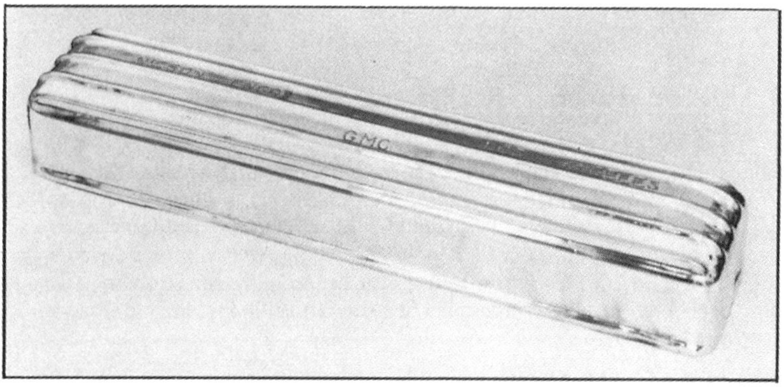


*Nicson heavy cast-aluminum rocker-arm cover for GMC. Similar available for all Chev. models 37-'51.*

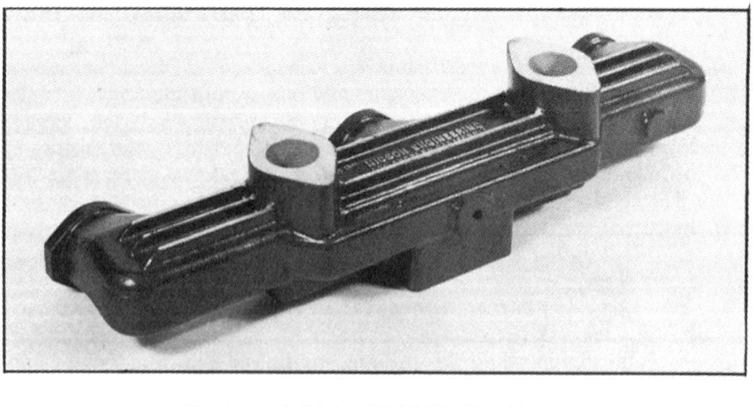

*Dual manifold for GMC 270 H (Nicson).*

*Nicson triple manifold with heat risers, Zenith carbs. Similar models available for Chev. and GMC without the heat risers.*

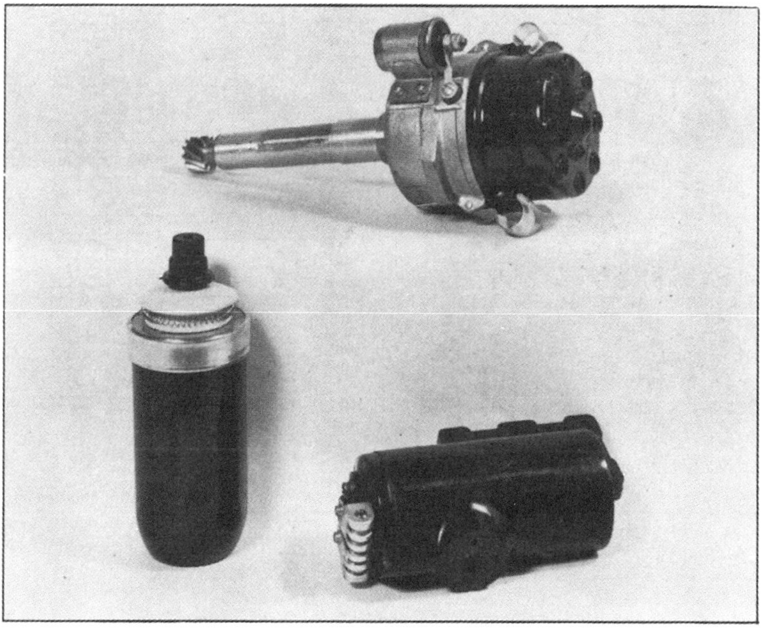

*Mallory dual-point distributor, special Mallory coil (left), and special DSM coil (right).*

# Decimal Equivalents

| Fraction | Decimal | Fraction | Decimal |
|---|---|---|---|
| 1/64 | .015625 | 33/64 | .515625 |
| 1/32 | .03125 | 17/32 | .53125 |
| 3/64 | .046875 | 35/64 | .546875 |
| 1/16 | .0625 | 9/16 | .5625 |
| 5/64 | .078125 | 37/64 | .578125 |
| 3/32 | .09375 | 19/32 | .59375 |
| 7/64 | .109375 | 39/64 | .609375 |
| 1/8 | .125 | 5/8 | .625 |
| 9/64 | .140625 | 41/64 | .640625 |
| 5/32 | .15625 | 21/32 | .65625 |
| 11/64 | .171875 | 43/64 | .671875 |
| 3/16 | .1875 | 11/16 | .6875 |
| 13/64 | .203125 | 45/64 | .703125 |
| 7/32 | .21875 | 23/32 | .71875 |
| 15/64 | .234375 | 47/64 | .734375 |
| 1/4 | .25 | 3/4 | .75 |
| 17/64 | .265625 | 49/64 | .765625 |
| 9/32 | .28125 | 25/32 | .78125 |
| 19/64 | .296875 | 51/64 | .796875 |
| 5/16 | .3125 | 13/16 | .8125 |
| 21/64 | .328125 | 53/64 | .828125 |
| 11/32 | .34375 | 27/32 | .84375 |
| 23/64 | .359375 | 55/64 | .859375 |
| 3/8 | .375 | 7/8 | .875 |
| 25/64 | .390625 | 57/64 | .890625 |
| 13/32 | .40625 | 29/32 | .90625 |
| 27/64 | .421875 | 59/64 | .921875 |
| 7/16 | .4375 | 15/16 | .9375 |
| 29/64 | .453125 | 61/64 | .953125 |
| 15/32 | .46875 | 31/32 | .96875 |
| 31/64 | .484375 | 63/64 | .984375 |
| 1/2 | .5 | 1 | 1. |

## EXPLANATION OF ENGINE CAPACITY

In Europe or Great Britain the displacement of an engine is usually given in liters (or "litres"), whereas in the United States displacement is generally referred to in terms of cubic inches. Below is table for converting liters to cubic inches:

| | | | |
|---|---|---|---|
| 1/2 | liter | 30.5125 | cu. in. |
| 1 | " | 61.025 | " " |
| 1 1/2 | " | 91.5375 | " " |
| 2 | " | 122.050 | " " |
| 2 1/2 | " | 152.5625 | " " |
| 3 | " | 183.073 | " " |
| 3 1/2 | " | 213.5855 | " " |
| 4 | " | 244.100 | " " |
| 4 1/2 | " | 274.6125 | " " |
| 5 | " | 305.125 | " " |
| 5 1/2 | " | 330.6375 | " " |
| 6 | " | 366.150 | " " |
| 6 1/2 | " | 396.6625 | " " |
| 7 | " | 427.175 | " " |

(From the above table it will be noted that Duesenberg, Miller, and Offenhauser engines were built to the "liter" standards; i.e.,

| | |
|---|---|
| 91 cu. in. | 1 1/2 liters |
| 183 " " | 3 " |
| 274 " " | 4 1/2 " |

## TIME — SPEED TABLE — OVER ONE MILE

| Time per Mile m. s. | Speed m.p.h. | Speed k.p.h. | Time per Mile m. s. | Speed m.p.h. | Speed k.p.h. | Time per Mile m. s. | Speed m.p.h. | Speed k.p.h. |
|---|---|---|---|---|---|---|---|---|
| 0 24 | 150.00 | 241.40 | 1 6 | 54.54 | 87.77 | 1 46 | 33.96 | 54.65 |
| 0 26 | 138.46 | 222.82 | 1 8 | 52.94 | 85.20 | 1 48 | 33.33 | 53.64 |
| 0 28 | 128.57 | 206.91 | 1 10 | 51.43 | 82.77 | 1 50 | 32.72 | 42.65 |
| 0 30 | 120.00 | 193.12 | 1 12 | 50.00 | 80.46 | 1 52 | 32.14 | 41.72 |
| 0 32 | 112.50 | 181.05 | 1 14 | 48.65 | 78.29 | 1 54 | 31.58 | 50.82 |
| 0 34 | 105.88 | 170.39 | 1 16 | 47.37 | 76.23 | 1 56 | 31.03 | 49.93 |
| 0 36 | 100.00 | 160.93 | 1 18 | 46.15 | 74.27 | 1 58 | 30.50 | 49.08 |
| 0 38 | 94.74 | 152.45 | 1 20 | 45.00 | 72.42 | 2 0 | 30.00 | 48.28 |
| 0 40 | 90.00 | 144.84 | 1 22 | 43.90 | 70.65 | 2 5 | 28.80 | 46.35 |
| 0 42 | 85.71 | 137.93 | 1 24 | 42.86 | 68.96 | 2 10 | 27.69 | 44.56 |
| 0 44 | 81.81 | 131.66 | 1 26 | 41.86 | 67.36 | 2 15 | 26.66 | 42.90 |
| 0 46 | 78.26 | 125.94 | 1 28 | 40.91 | 65.84 | 2 20 | 25.71 | 41.37 |
| 0 48 | 75.00 | 120.70 | 1 30 | 40.00 | 64.37 | 2 25 | 24.83 | 39.96 |
| 0 50 | 72.00 | 115.87 | 1 32 | 39.13 | 62.97 | 2 30 | 24.00 | 38.62 |
| 0 52 | 69.23 | 111.41 | 1 34 | 38.29 | 61.62 | 2 35 | 23.32 | 37.37 |
| 0 54 | 66.66 | 107.28 | 1 36 | 37.54 | 60.35 | 2 40 | 22.50 | 36.21 |
| 0 56 | 64.28 | 103.45 | 1 38 | 36.73 | 59.11 | 2 45 | 21.81 | 35.10 |
| 0 58 | 62.07 | 99.89 | 1 40 | 36.00 | 57.93 | 2 50 | 21.17 | 34.07 |
| 1 0 | 60.00 | 96.56 | 1 42 | 35.29 | 56.79 | 2 55 | 20.57 | 33.10 |
| 1 2 | 58.06 | 93.44 | 1 44 | 34.61 | 55.70 | 3 0 | 20.00 | 32.18 |
| 1 4 | 56.25 | 90.52 | | | | | | |

*Note*—Comparative Distances in Miles and Kilometres may also be found by this table, *e.g.* 60.00 miles equal 96.56 kilometres.

## SPEED — TIMETABLE — OVER TENTHS of a MILE

| Tenths of a Mile | 20 m.p.h. | 21 m.p.h. | 22 m.p.h. | 23 m.p.h. | 24 m.p.h. | 25 m.p.h. | 26 m.p.h. | 27 m.p.h. | 28 m.p.h. | 29 m.p.h. | 30 m.p.h. |
|---|---|---|---|---|---|---|---|---|---|---|---|
| | Time in Seconds. | | | | | | | | | | |
| .1 | 18 | 17.1 | 16.4 | 15.7 | 15 | 14.4 | 13.9 | 13.3 | 12.9 | 12.4 | 12 |
| .2 | 36 | 34.2 | 32.7 | 31.3 | 30 | 28.8 | 27.7 | 26.7 | 25.7 | 24.8 | 24 |
| .3 | 54 | 51.4 | 49.1 | 47.0 | 45 | 43.2 | 41.6 | 40.0 | 38.6 | 37.2 | 36 |
| .4 | 72 | 68.6 | 65.4 | 62.6 | 60 | 57.6 | 55.4 | 53.3 | 51.4 | 49.6 | 48 |
| .5 | 90 | 85.7 | 81.8 | 78.3 | 75 | 72.0 | 69.3 | 66.6 | 64.3 | 62.1 | 60 |
| .6 | 108 | 102.8 | 98.2 | 93.9 | 90 | 86.4 | 83.1 | 80.0 | 77.2 | 74.5 | 72 |
| .7 | 126 | 119.9 | 114.5 | 109.6 | 105 | 100.8 | 97.0 | 93.3 | 90.0 | 86.9 | 84 |
| .8 | 144 | 137.1 | 130.9 | 125.2 | 120 | 115.2 | 110.8 | 106.6 | 102.9 | 99.3 | 96 |
| .9 | 162 | 154.3 | 147 2 | 140 9 | 135 | 129.6 | 124.7 | 120.0 | 115.7 | 111.7 | 108 |
| .0 | 180 | 171.4 | 163.6 | 156.5 | 150 | 144.0 | 138.5 | 133.3 | 128.6 | 124.1 | 120 |

*Note.*—This table will be found useful both in checking speedometer readings and in those trials where a secret check necessitates an exact average speed.

## TIME — SPEED TABLE — OVER QUARTER MILE

| Sec. | M.p.h. | Sec. | M.p.h. | Sec. | M.p.h. | Sec. | M.p.h. | Sec. | M.p.h. | Sec. | M.p.h. |
|---|---|---|---|---|---|---|---|---|---|---|---|
| 15 | 60.00 | 13 4/5 | 65.22 | 12 3/5 | 71.43 | 11 2/5 | 78.95 | 10 1/5 | 88 24 | 9 | 100.00 |
| 14 4/5 | 60.81 | 13 3/5 | 66.18 | 12 2/5 | 72.58 | 11 1/5 | 80.36 | 10 | 90.00 | 8 4/5 | 102.47 |
| 14 3/5 | 61.64 | 13 2/5 | 67.16 | 12 1/5 | 73.77 | 11 | 81.82 | 9 4/5 | 91.84 | 8 3/5 | 104.65 |
| 14 2/5 | 62.50 | 13 1/5 | 68.18 | 12 | 75.00 | 10 4/5 | 83.33 | 9 3/5 | 93.75 | 8 2/5 | 107.14 |
| 14 1/5 | 63.38 | 13 | 69.23 | 11 4/5 | 76.27 | 10 3/5 | 84.91 | 9 2/5 | 95.74 | 8 1/5 | 109.76 |
| 14 | 64.28 | 12 4/5 | 70.31 | 11 3/5 | 77.59 | 10 2/5 | 86.54 | 9 1/5 | 97.83 | 8 | 112.50 |

**From INTERNATIONAL SPORTING CALENDAR, 1951,**

**By Brittains Garage**

# About Floyd Clymer

from the original inside cover of *Souping the Stock Engine:*

Floyd Clymer grew up with the American automobile and is a man well qualified to evoke memories of Barney Oldfield, the Stutz Bearcat, the Stanley Steamer, the Glidden Tours, the Octoauto (eight wheels), the Duck (with its back-seat steering wheel), the Tin Lizzie, and that original Horseless Carriage which mounted a life-size horse's head on the radiator. For half a century, Floyd Clymer has tested, raced and restored American automobiles. He has even invented accessories for them. Teddy Roosevelt called him the "the world's youngest automobile salesman" and he once held the motorcycle speed record to the top of Pike's Peak in Colorado. In recent years he has written and published a score of books on the automobile.

Floyd Clymer also published the 1954 California Bill Chevrolet, GMC & Buick Speed Manual and the 1952 California Bill Ford Speed Manual when Bill Fisher closed that business in the mid-50s. These two California Bill publications have also been reproduced and are again available through Fisher Books and its dealers.

*Disclaimer*

When this book was originally published, Floyd Clymer or Roger Huntington would never have thought to include a disclaimer. But in today's society it is important to include a disclaimer in anything that is published.

This information is republished as a fun project to make history available to auto enthusiasts. The material in this book was created around 1949. It is offered solely as an indication of how automotive performance was obtained at back then.

The parts, materials, processes or anything else shown or discussed in he book will probably not be available in the 1990s. There's no guarantee that the methods and materials presented herein will be suitable for the engine modifications or racing that you might want to do today. Over 46 years of development have changed many concepts and methods since this book was first published in 1951. The general ideas of the paths to power are still totally valid, but the specifics and recommendations may have changed in many instances.

Many of the manufacturers mentioned are no longer in business.
Most of the parts that are shown and described have not been made for many years. If you can find them they may only be available at very high prices at swap meets or through automotive enthusiasts' magazine ads, such as in Hemmings.

# *Other nostalgia books you'll be interested in:*

**$14.95 pb ISBN 1-55561-137-0**
5.5 x 8.5, 192 pages, 1950 Edition

## *Souping the Stock Engine*

Roger Huntington

Reprint of the 1950 classic, this true hotrodders' guidebook provides effective methods for tuning all types of stock engines from the conservative road car to full-race capability. Includes general engine performance, stock-engine characteristics, and basic planning for modifications.

**$14.95 pb • ISBN 1-55561-106-0**
5.5 x 8.5, 128 pgs,125 b&w photos & illus.
2nd printing

## *Chevrolet, GMC & Buick Speed Manual*

Bill Fisher

Reprint of original 1952 edition. How to hotrod Chevrolet inline six-cylinder 216 and 235 cubic-inch engines, plus GMC 228, 248, 256, 270 and 302-CID engines, and the Buick straight-eight 248 and 320-CID engines. This is how it was done in the 50s!

***At your bookstore or specialty automotive store.***